Lecture Notes in Artificial Intelligence   1152

Subseries of Lecture Notes in Computer Science
Edited by J. G. Carbonell and J. Siekmann

Lecture Notes in Computer Science

Edited by G. Goos, J. Hartmanis and J. van Leeuwen

Springer

Berlin
Heidelberg
New York
Barcelona
Budapest
Hong Kong
London
Milan
Paris
Santa Clara
Singapore
Tokyo

Takeshi Furuhashi   Yoshiki Uchikawa (Eds.)

Fuzzy Logic, Neural Networks, and Evolutionary Computation

IEEE/Nagoya-University
World Wisepersons Workshop
Nagoya, Japan, November 14-15, 1995
Selected Papers

Springer

Series Editors

Jaime G. Carbonell, Carnegie Mellon University, Pittsburgh, PA, USA
Jörg Siekmann, University of Saarland, Saarbrücken, Germany

Volume Editors

Takeshi Furuhashi
Yoshiki Uchikawa
Nagoya University, Department of Information Electronics
Furo-cho, Chikusa-ku, Nagoya 464-01, Japan
E-mail: {furuhashi/uchikawa}@nuee.nagoya-u.ac.jp

Cataloging-in-Publication Data applied for

Die Deutsche Bibliothek - CIP-Einheitsaufnahme

Fuzzy logic, neural networks and evolutionary computation :
selected papers / IEEE/Nagoya-University World Wisepersons
Workshop, Nagoya, Japan, November 14 - 15, 1995. Takeshi
Furuhashi ; Yoshiki Uchikawa (ed.). - Berlin ; Heidelberg ;
New York ; Barcelona ; Budapest ; Hong Kong ; London ;
Milan ; Paris ; Santa Clara ; Singapore ; Tokyo : Springer, 1996
  (Lecture notes in computer science ; Vol. 1152 : Lecture notes in
  artificial intelligence)
  ISBN 3-540-61988-7
NE: Furuhashi, Takeshi [Hrsg.]; World Wisepersons Workshop <4, 1995,
  Nagoya>; Industrial Electronics Society; GT

CR Subject Classification (1991): I.2, F.1-2

ISBN 3-540-61988-7 Springer-Verlag Berlin Heidelberg New York

© Springer-Verlag Berlin Heidelberg 1996
Printed in Germany

Typesetting: Camera ready by author
SPIN 10513801   06/3142 – 5 4 3 2 1 0   Printed on acid-free paper

Preface

Fuzzy logic is one of the key technologies for representing human knowledge in the brain and for constructing adaptive systems. Difficulties for fuzzy logic include knowledge acquisition from experts and/or knowledge finding for unknown tasks. Neural networks (NNs) and evolutionary computation (EC) are also gaining attention for their potential for knowledge representation/finding. Combining the technologies of fuzzy logic and neural networks/evolutionary computation is expected to open a new paradigm in machine learning for the realization of human-like information generating system. This is the background against which the IEEE/Nagoya University World Wisepersons Workshop (WWW'95) on Fuzzy Logic and Neural Networks/ Evolutionary Computation was organized.

This workshop is the fourth of the WWW series organized to cover the rapidly advancing field of research on the combined technologies of Fuzzy - NN and Fuzzy - EC. IEEE/Nagoya University WWW is a series of workshops sponsored by Nagoya University and co-sponsored by the IEEE Industrial Electronics Society. The third one (WWW'94) was organized by Associate Prof. Furuhashi (Dept. of Information Electronics, Nagoya University) in August 1994 to allow attendees to have a lively discussion on the subject "Present and Future Situation of the Combination Technologies of Fuzzy Logic and Neural Networks/Genetic Algorithm", an area which has made remarkable progress recently. The selected papers of WWW'94 were published as LNAI/LNCS Vol. 1011 in 1995.

The fourth WWW is to extend and deepen the discussions from WWW'94. The steering committee has accepted 17 excellent papers. As the editors for the selected papers of WWW'95, we selected twelve excellent papers which were revised, extended, and rigorously reviewed by three reviewers each.

The first five papers discuss interesting and advancing topic on the combined technology Fuzzy - EC. M. A. Lee, H. Esbensen, and L. Lemaitre develop techniques to design mutiobjective optimization algorithms based on hybrid fuzzy system/evolutionary algorithm techniques. The technique is based on pioneering work done by one of the authors which uses a fuzzy system to control the evolutionary algorithm. S. Nakanishi et al. propose a new fuzzy modeling method using the genetic algorithm. Structure identification of the fuzzy models has been one of the difficult tasks of the fuzzy modeling. Hierarchical structures of unknown objects can be identified from the input and output data by the proposed method. H. Ishibuchi, T. Nakashima, and T. Murata present a fuzzy classifier system for automatic generation of a compact fuzzy system based on a small number of linguistic classification rules. T. Nomura and T. Miyoshi propose a new genetic operator called Unfair Average Crossover for fuzzy clustering using real value coding. The parameters of membership functions of fuzzy if-then rules are directly encoded. T. Yoshikawa et al. present a new coding method using artificial DNA based on an analogy with biological DNA. A new mechanism of development from the artificial DNA is also presented. Redundancy and overlaps of genes are the unique features of the proposed method. Fuzzy rules are found by the proposed combination technology of the new method.

In the sixth paper, T. Aoki et al. propose a new architecture consisting of hierarchical fuzzy rules, a fuzzy performance evaluation system, and learning automata. The evaluation function of the learning automata is constructed using fuzzy logic for coping with a dynamically changing environment.

The next three papers discuss hybrid technology of Fuzzy - NN. The seventh paper discusses a timely and important problem of fuzzy modeling. P. J. Costa Branco, N. Lori, and J. A. Dente study two pre-processing methods for structure identification of fuzzy models: a statistical method called Principal Component Analysis and a clustering technique called Autonomous Mountain-Clustering Method. These two methods are effective for efficient fuzzy modeling. with neuro-fuzzy techniques. The eighth paper, by A. Z. Kouzani and A. Bouzerdoum, introduces a generic fuzzy neuron. A fuzzy neural network architecture based on the generic fuzzy neuron is developed for motion estimation. A. Buller presents a unique Fizzy-Fuzzy inference method utilizing a working memory inhabited by a society of Running Agents (RATs). This method is devised for fuzzy knowledge processing in a massively parallel way.

The last three papers also tackle interesting problems. The tenth paper, presented by E. Perez-Minana, P. Ross and J Hallam, describes multi-layer perceptron design utilizing the information subsumed in the Delaunay Triangulation and Voronoi Diagram for achieving a reasonable network for the task of interest. M. Inuiguchi et al. successfully applied a genetic algorithm to an actual problem: coal purchase planning in a real electric power plant. The last paper, by H. de Garis, reports a challenging project, the CAM-Brain Project, which aims to use evolutionary engineering techniques to build/grow/evolve a RAM-and-cellular-automata based artificial brain consisting of thousands of interconnected neural network modules inside special hardware. I am now proud to have these papers published in the LNCS/LNAI series.

Takeshi Furuhashi
Yoshiki Uchikawa

Nagoya, July 1996

Contents

The Design of Hybrid Fuzzy/Evolutionary Multiobjective Optimization Algorithms

Michael A. LEE   Henrik ESBENSEN   Laurent LEMAITRE[*]
Department of Electrical Engineering and Computer Sciences
University of California, Berkeley, CA 94720 USA
leem@cs.berkeley.edu esbensen@eecs.berkeley.edu lemaitre@geneve.sps.mot.com
fax: 510 642 5775

Abstract. In this paper, we develop and demonstrate techniques to design multiobjective optimization algorithms based on hybrid fuzzy system/evolutionary algorithm techniques. The technique is based on an approach where a fuzzy system is used to control the evolutionary algorithm. By viewing the search process as a dynamic process, high performance strategies are developed using controller design techniques. Through the use of indicators aimed at assessing the performance of evolutionary algorithms for multiobjective optimization, we show how to design fuzzy systems for controlling the search behavior. The key contributions of the work reported in this paper are the techniques for quantitatively measuring the performance of population based multiobjective optimization algorithms and techniques for automatically designing optimization algorithms. We demonstrate our techniques on an Integrated Circuit placement task that includes timing and geometrical objectives.

1 Introduction

In many real-world contexts, humans are forced to choose among solutions that vary in performance with respect to multiple competing objectives. For example, in the case of integrated circuit (IC) layout generation, geometrical and timing specifications interact and performance along each of these directions may need to be compromised to satisfy specifications [4]. In problems of this nature often more than one solution exists, each of which must be regarded as equal in the absence of information relating the relative importance between the competing objectives. The final solution chosen is ultimately based on the specifications of the application or subjective (i.e. marketing) criteria, which can be difficult to define.

One approach to this problem is known as multiobjective optimization. The goal of this approach is not to provide the users with a single solution, but rather a set of solutions that represent the set of best alternatives, or *Pareto Optimal* set. The notion of Pareto optimality is based on the concept of dominance. Let $f(x) = (f_1(x), ..., f_n(x))$ represent a vector valued objective function and u and v represent two solutions. u

---

[*] This work was initiated when the author was a Siemens Visiting Industrial Fellow at UC Berkeley. The author is now with Motorola in Geneve, Switzerland.

dominates v, written $u <_d v$, if and only if $(\forall i: f_i(u) \le f_i(v)) \wedge (\exists i: f_i(u) < f_i(v))$. Solutions included in the *Pareto Optimal Set* are those that cannot be improved along any dimension without simultaneously being deteriorated along other dimension(s).

In this paper, we present an evolutionary algorithm approach to multiobjective optimization. These techniques have been shown to work well on problems such as gas turbine engine design and fuzzy systems design. However, these algorithms do not always behave as we expect or in an intuitively efficient manner. Therefore, we propose using a dynamic control framework to modify the behavior of these algorithms. In particular, we will present techniques to control multiobjective evolutionary algorithms using a fuzzy systems approach. The following sections will review multiobjective evolutionary algorithms, introduce the control framework for the algorithms, present techniques to design control strategies. The last three sections will present some results obtained from using our hybrid algorithms on the IC placement problem and conclusions.

2 Multiobjective Evolutionary Algorithms

Evolutionary algorithms are population based stochastic search strategies modeled after natural genetics mechanisms [8]. An individual of a population encodes a solution as a string of parameters, which is then subjected to genetic operations such as mutation and crossover operations. An individual's likelihood to pass information onto the next generation is determined by its ability to thrive in the target environment: evolutionary algorithms implement a *survival of the fittest* policy.

Using a simple evolutionary algorithm entails three main steps:

- design a solution representation
- design a genetic encoding of the solution
- design an appropriate evaluation function

In practice, virtually all evolutionary algorithms use a single valued function to drive the selection; that is, the performance of an individual is aggregated into a single value. However, as mentioned in the introductory section, it can be difficult to combine satisfactorily the objectives into a single value.

3 Hybrid Fuzzy-Genetic Algorithms

In our work, we extend hybrid fuzzy system/evolutionary algorithm techniques developed in [10,11]. The previously developed approach used a fuzzy system to monitor and control the behavior of a genetic algorithm. A schematic diagram of the system is shown in Figure 1. An example fuzzy system can include rules that relate indicators measuring diversity or distribution characteristics to control parameters such as population size or mutation rate. A detailed view of the inputs and outputs to a typical fuzzy system in this context is shown in Figure 2. The inputs and outputs of this controller are only suggestions as different indicators and control outputs can be used. This

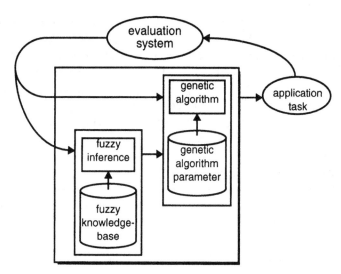

Figure 1. Schematic diagram of the Dynamic Parametric Genetic Algorithm. Genetic algorithm search behavior is monitored and controlled by a fuzzy system.

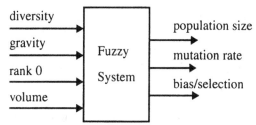

Figure 2. Inputs and outputs to a typical fuzzy system used to monitor and control a multiobjective evolutionary algorithm. Details of the particular inputs and outputs of this system are discussed in the text.

approach has been shown to improve search performance of single objective optimization algorithms. This paper aims at applying these techniques to multiobjective optimization algorithms.

The next subsection focuses on developing indicators that characterize the search process and use them as inputs to a fuzzy system that effects various parameters of the search algorithm, such as selection operator parameters. The following subsection describes a meta-level evolutionary algorithm technique for automatically designing a fuzzy system to control the multiobjective evolutionary algorithm.

3.1 Characterizing Solution Sets

Search performance is difficult to define and can be dependent on the final application; what constitutes a good search depends on the quality of the output within a given context. However, once some performance measures are defined, the search performance can be optimized using techniques such as those proposed in [7] and [11].

Several generic search performance measures have been invented such as online and offline measures [3]. Online measure is a running average of all solutions evaluated up to a given point in time and offline measure is a running average of the best solution evaluated up to a given point in time. These measures have been used to tune genetic algorithm parameters and develop fuzzy system controllers to improve the performance of single objective optimization algorithms. As a result they must be adapted for use in multiobjective optimization applications. A discussion of the modification of these measures will be given below.

In the case of multiobjective optimization, the goal is to provide the users with a set of solutions to choose among. The following properties of the final solution set, in the objective space $f(X)$, may indicate how well the Pareto optimal set is represented (for minimization):

- the diversity of the non-dominated solutions in the final set is maximized
- the number of non-dominated solutions in the final set is maximized
- the bounding volume of the non-dominated solutions set is maximized
- the center of gravity of the final solution set is close to the origin

These properties correspond to moving the cloud of points toward the origin, but also spreading out uniformly, in a sense, to cover the true Pareto optimal set. The pressure to pull the center of gravity toward the origin corresponds to simultaneous minimization of all constraints. In the remainder of this section, we will define the proposed indicators.

Let $r(y, X) = \left| \{ x \in X | (x <_d y) \} \right|$ be the rank of solution y with respect to the set X. Then the set of non-dominated solutions, X_0, can be defined as: $X_0 = \{ x \in X | r(x, X) = 0 \} \subseteq X$.

The diversity measure of the non-dominated solutions can be defined as $d(X) = \frac{1}{|X_d|} \sum_{(x, y) \in X_d} dist(f(x), f(y))$ where $X_d = \{ (x, y) \in X_0^2 \,|\, x \neq y \}$ and $dist$ is a distance measure in the cost space.

Maximizing the number of non-dominated solutions in the final set forces a more representative sampling of the Pareto optimal set. For this purpose we maximize $c(x) = 1 + \sum_{x \in X} r(x, X)$.

Maximizing the bounding volume has the effect of ensuring optimization along individual objectives and covering the true Pareto optimal set. The bounding volume is defined as $v(X) = \prod_{i=1}^{n} max_{x \in X_0} \{ f_i(x) \}$.

Center of gravity is defined as $g(X) = dist(\vec{0}, \bar{g}(X))$, where $\bar{g}_i$, the *ith* component of the center of mass, is defined as $\bar{g}_i(X) = \frac{1}{|X|} \sum_{x \in X} f_i(x)$.

3.1.1 Characterizing Dynamic Search Behavior

In single objective evolutionary algorithms, it is common practice to store the best solution found either external or internal to the population. In a multiple objective optimization setting, the final result desired is not just one solution, but a family of solutions. The analogous procedure is to store the non-dominated solutions. There is an extra operation when merging the non-dominated solutions of the current population with the saved set, which is to remove any solutions that are replaced by incoming solutions. That is, the non-dominated set determination must be done on the saved set. When the evolutionary algorithm terminates, we are left with a set that, for our particular run, best approximates the true Pareto optimal set.

In our work, we are interested in controlling the behavior of search algorithms such that they exhibit high performance. One way to approach this problem is to treat the combination of a search process and a performance measure as a function to optimize. However, most optimization algorithms are one dimensional and therefore the multiple indicators we presented must be combined. One of the simplest methods to combine the indicators is:

$$q(X) = \frac{v(X)d(X)}{g(X)c(X)}. \tag{1}$$

This measure only represents the quality of a single set. Because we are interested in the dynamic behavior of search algorithms, we present the following dynamic set quality measures:

$$online_{mo} = \frac{\sum_{t=0}^{T} q(X^t)}{T+1} \tag{2}$$

$$offline_{mo} = \frac{\sum_{t=0}^{T} q(X_0^t)}{T+1} \tag{3}$$

where X^t is the set at time t. These measures give us an indication of how the set quality changes over time. Both of these measures should be maximized when the target algorithm performs minimization on all objective dimensions. (These measures can be seen as generalizations of performance measures used to compare single objective search algorithms [3].)

3.2 Designing Search Control Strategies

One approach to designing the fuzzy system used to monitor and control the search behavior is a meta-level technique. The task at hand is that of designing a fuzzy controller for a dynamic process. Many techniques such as neuro-fuzzy and evolutionary techniques for adapting fuzzy system exist. In this work, we will use a fuzzy system design technique proposed in [12].

The fuzzy system is coded as a string of information which is then searched using a meta-level genetic algorithm. In this case, the fitness of a fuzzy system will be how well it is able to control a search algorithm to maximize a search performance measure. For example, we can use the offline measure formulated in (3) and search for a fuzzy system that maximizes this measure when the algorithm is run on several test functions (where the test functions represent the types of problems we are ultimately interested in). A diagram showing the relationship between the meta-level search and evolutionary search algorithm is shown in Figure 3.

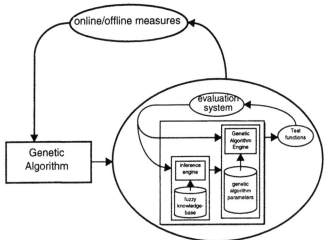

Figure 3. Meta-level technique for evolving hybrid fuzzy-genetic multiobjective evolutionary algorithms. Algorithms are optimized with respect to search performance measures such as the online and offline measures formulated in eq. 2 and 3 in the text. The test problems are representative of the class of problems of ultimate interest.

In the following sections, we describe the system and genetic representations of the fuzzy controller.

3.2.1 Shared-Triangular Fuzzy System Architecture

The *Shared-Triangular* representation uses asymmetrical triangular membership functions and the *min* operator to synthesize multidimensional membership functions[9]. Each triangular membership function is specified by its center, left base width, and right base width.

Rules cover the input space by selecting and combining the one-dimensional membership functions from a globally defined set (all rules have access to the same set of membership functions). There is also a possibility for a rule to have no membership function associated with a particular input dimension, which implicitly forces the rule to cover entirely an input dimension.

Each rule has an additional parameter to modify, or hedge, the rule firing strength. If we define $\mu(\hat{x})$ as the raw rule firing strength, then the modified firing strength of a rule is given by:

$$\mu^p(\hat{x}),\qquad(4)$$

where p represents the degree of hedging. The consequent output of each rule has a Takagi-Sugeno-Kang (TSK) form with the output of the system computed by taking a weighted sum of the outputs of all firing rules[12].

3.2.2 Genetic Representation

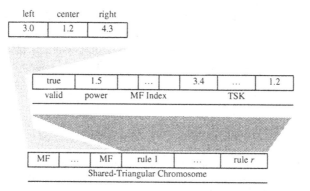

Figure 4. Shared-Triangular genetic representation showing rule validity parameter and rule power parameter. Note that there are $b + 1$ TSK parameters.

The genetic representation for the Shared-Triangular fuzzy systems is shown in Figure 4. The genetic code is made up of two different macro structures: membership function genes and rule genes. The membership genes specify distance between adjacent triangle centers and the left and right triangle base widths [10]. The rule genes have locations for the membership function set indices for each input dimension as well as a validity bit. There is also a provision for completely covering an input dimension when the index is set to a special value.

The total number of system parameters in an b-dimensional system with r rules and a maximum of c membership functions per input dimension is:

$$3bc + r(2b + 3). \tag{5}$$

Each rule requires $2b + 3$ parameters, because the membership function associated with each dimension must be specified (we implement this via an indexing scheme).

4 Experiments and Results

The techniques presented in this paper have been investigated using the application briefly described below as a case study. For a detailed description of the application the reader is referred to [4].

4.1 The IC Placement Problem

A certain type of integrated circuits (ICs) consist of a number of interconnected blocks. Each block is rectangular and contains a number of interconnected transistors implementing some part of the functionality of the circuit. For example, in a CPU, RAM and ALUs constitute blocks. The IC placement problem is that of placing the blocks in the plane such that a number of competing objectives are optimized while satisfying given geometrical constraints, of which the most important is that blocks cannot overlap. Communication between blocks is physically made possible by (a large number of) wires connecting specific points within each block to points within other blocks. Since the locations of connection points within a block moves as the block is rotated and/or mirrored in one or both of the axes, determining which one of eight possible orientations for each block is also part of the placement problem. A sample placement is shown in Figure 5.

The algorithm in [4] minimizes three competing criteria:

- The layout area, i.e., the area of the smallest bounding box containing all

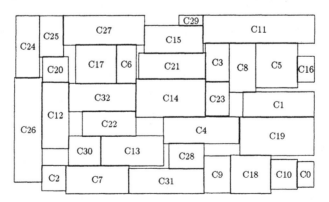

Figure 5. A layout of 33 blocks named C0 through C32. No wires or connection points are shown.

blocks. The area is the main factor determining the production cost of the circuit.

- The deviation of the aspect ratio of the layout (height divided by width) from a user-defined target value. The aspect ratio has to be within a certain range in order for the layout to fit in a given circuit package.
- The maximum estimated time it may take some signal in the circuit to propagate from one memory element to another, passing through a sequence of blocks an interconnecting wires. Minimizing this quantity is equivalent to maximizing the clock frequency at which the circuit will be able to operate correctly.

The placement problem is of significant interest to the IC design industry, and is very challenging due to its complexity. It is NP-hard and the search spaces represented by practical problem instances are usually extremely large. Furthermore, the criteria considered, especially the clock frequency maximization, are ill-behaved functions of the block placement. Consequently, high-performance state-of-the-art circuits are still being placed manually.

In [4], a *goal vector* is specified by the user, defining a goal value for each of the three criteria optimized. A preference relation between solutions, incorporating the goal vector, is then applied to search for a set of good solutions relative to the goals, as introduced in [6]. However, for simplicity, the goal vector is not used in this work. The preference relation used then reduces to the notion of dominance, and the search performed is the search for a sample of the Pareto optimal set (Section 1).

4.2 Algorithm Characteristics

The placement algorithm is a steady-state, hybrid GA, making extensive use of problem-specific knowledge. The essential part of the phenotype is an inverse Polish expression specifying the relative positions of all blocks. Only placements known as *slicing structures* are considered, for which a unique representation by Polish expressions were first introduced in [14]. A placement is a slicing structure if it is possible to recursively partition the placement into two parts by a cut-line such that, in the end, each partition contains only one block. Each cut-line has to be horizontal or vertical, span the entire layout, and not intersect any block. Given a slicing structure, the absolute positions and orientations of blocks are determined using problem-specific algorithms. Blocks are oriented by an algorithm assuring that the minimum area for the given Polish expression is obtained, and a compaction algorithm is then applied to further reduce layout area. The phenotype and its interpretation is illustrated in Figure 6.

Crossover and mutation of the Polish expression parts of the phenotypes are performed using operators introduced in [2]. A key feature of these operators is that they all preserve feasibility, that is, only feasible solutions, which can be interpreted as valid placements using the scheme outlined above, are ever considered by the GA.

Selection for crossover is rank based following [6,13], i.e., no explicit fitness computa-

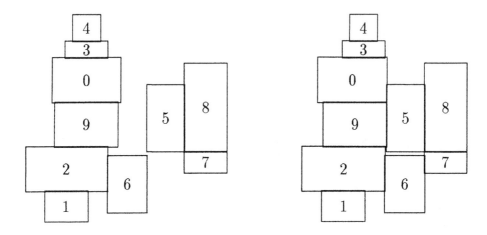

Figure 6. Given 10 blocks and the Polish expression *1 2 + 6 * 9 0 + + 3 4 + + 5 * 7 8 + ***, where + and * represent a vertical and horizontal cut-line, respectively, and the numbers are block identities. The placement to the left is the result of placing and recursively centering subtrees of the expression while orienting each block so that the minimum layout area is obtained. Further reduction of the layout area is then attempted using a compaction algorithm, yielding the placement to the right, which is the genotype placement.

tion is performed. Given a population $\Phi = \{\phi_0, ..., \phi_{M-1}\}$, and assume $r(\phi_0, \Phi) \leq ... \leq r(\phi_{M-1}, \Phi)$. Each parent ϕ is selected such that: 1) the probability that $r(\phi, \Phi)$ equals $r(\phi_k, \Phi)$, $P[r(\phi, \Phi) = r(\phi_k, \Phi)]$, decreases linearly with k, and $P[r(\phi, \Phi) = r(\phi_0, \Phi)] = \beta P[r(\phi, \Phi) = r(\phi_{M/2}, \Phi)]$, where a user-defined parameter, $1 < \beta \leq 2$, controls the selection pressure. 2) all individuals having the same rank have the same probability of being selected.

Generated offspring replace poor solutions, yielding an elitist strategy in which a currently non-dominated solution ϕ is never removed from the population unless its replacement dominates it.

4.3 Previous Results

It is inherently difficult to fairly compare an algorithm performing multi-dimensional optimization to a traditional one-dimensional optimization, e.g., minimizing a weighted sum of the criteria considered. However, in special cases, where a comparison to simulated annealing (SA) can be performed, the results obtained by the GA are shown in [4] to be comparable to those of SA, and in this sense, the GA approach to placement is very promising.

However, there are a number of unsolved problems not addressed in [4] relating to the solution sets generated. In some runs, the non-dominated solutions of the output set are very few and/or clustered in a small region of the cost-space, only providing the user with limited trade-off information. Hence, the need for developing solution set quality measures and systematic techniques for controlling and improving quality of the final solution set.

4.4 Experimental Method

As a case study, we have conducted experiments to validate our proposal to dynamically control parameters of the genetic algorithm. As in Figure 2, we monitor the genetic algorithm behavior using the indicators developed in Section 3.1. The cost dimensions are normalized before the indicators are computed. Outputs of our system are the change in population size, mutation rate, and the selection parameter, β. Each of the outputs was coded as a multiplier to the current value, i.e. $M_{new} = M_{old} \times \Delta M$. In addition, following the parameter update, hard limiting was enforced to bound population size to [10, 200], mutation rate to [0.00001, 0.01], and β to (1, 2.0]. The control actions were performed following the generation of M offspring.

We design a fuzzy system using a meta-level genetic algorithm technique that optimizes against the offline measure formulated in (3). The system representation of the fuzzy controller was such that the number of rules, membership function shapes, and consequent parts were searched at the same time [10]. The parameters of the meta-level genetic algorithm were set as: population size of 20, mutation rate of 0.01, and crossover rate of 0.6. The genetic algorithm also had various mechanisms such as windowing and generation gap.

4.5 Experimental Results

Figure 7-Figure 9 show typical behaviors for a random walk, static genetic algorithm, and a dynamic genetic algorithm designed against the offline measure. The random walk algorithm generates random solutions and updates and stores the set of non-dominated solutions. To equalize the search space size, the random walk generated random genetic strings which are passed through the same decoding and evaluation routines that the genetic algorithms use. The static genetic algorithm uses static parameter settings for population size, mutation rate, and bias. These values were set according to experience.

The static genetic algorithm performs as described in Section 4.2 (the parameters are static and the algorithm has not been optimized according to the offline performance

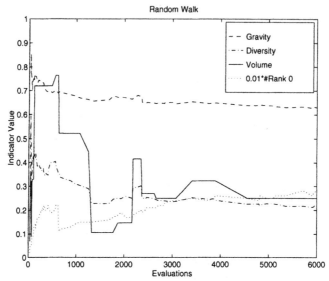

Figure 7. Dynamic behavior of random walk algorithm.

measure). The dynamic genetic algorithm was obtained using the technique outlined in the previous section.

The gravity measure of the static genetic algorithm reduces much more than in the random walk search, however the volume and diversity of the non-dominated solutions in the random walk are much greater than that of the static genetic algorithm's produced

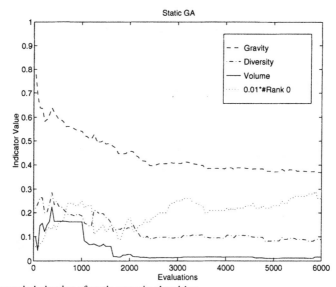

Figure 8. Dynamic behavior of static genetic algorithm.

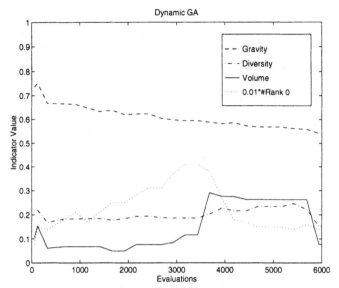

Figure 9. Dynamic behavior of dynamic genetic algorithm.

solution set. The behavior of the genetic algorithm corresponds to a much more exploitation oriented search; new solutions are more likely to be produced near existing good ones.

As shown in Figure 10, the dynamic genetic algorithm obtains a better offline behavior than the static genetic algorithm. By comparing Figures 8 and 9 it can be seen that the improvement is mainly caused by an increased volume.

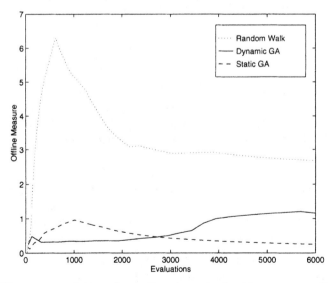

Figure 10. Offline measure of the random walk and genetic algorithms corresponding to data in Figure 7-Figure 9.

Furthermore, as expected, it was observed that both genetic algorithms performs significantly better than the random walk: The solution sets obtained by the random walk are always dominated by the sets obtained by any of the genetic algorithms. Nevertheless, the random walk has a much better offline performance than both genetic algorithms, see Figure 10. The reason is that the improved volume and diversity of the random walk has a greater impact on the offline measure than the inferior gravity.

Hence, while the technique used to design algorithms against a performance measure is successful (the dynamic algorithm outperforms its static counterpart), the measure itself fails to appropriately reflect our subjective notion of algorithm performance.

Since each of the set quality indicators introduced in Section 3.1 seems intuitively reasonable, the main problem must be the proper combination of the indicators into a single valued quantity expressing the overall set quality.

Current work aiming at solving this problem by redefining the solution set quality measure is briefly described in the following Section.

5 Solution Set Quality Revisited

The definition of solution set quality introduced and applied in the previous sections is based on explicit definition and combination of a number of individual set quality indicators.

A fundamentally different approach is to define solution set quality implicitly by modeling the choice process ultimately performed by the user of the algorithm, as described in Section 5.1. Preliminary results using the new measure are presented in Section 5.2.

5.1 An Alternative Quality Measure

From one or more given sets of solutions the user will ultimately select a single solution as the ``best''. For example, in the case of IC placement, a single layout will ultimately be selected for production. However, it is not known *how* the user makes the final choice; it depends on preferences which are probably never formally expressed.

The basic idea of the proposed alternative solution set quality measure is to model the final selection performed by the user by a selection function parameterized to account for different possible preferences with respect to the relative importance of the optimization criteria. By systematically varying the parameters of the selection function, a class of functions corresponding to a wide range of possible user-preferences is obtained. The quality $q(X)$ of a set X is then defined as the expected value of the selection function when selecting from X while traversing the class of selection functions.

The measure q is introduced and described in detail in [5]. Here we just outline the basic definition and its properties. As in Section 4.4 a normalization of the cost dimensions are still needed. Let $\overline{f}_i(x) \in [0, 1]$ denote the normalized cost of $x \in X$ with respect to the $i'th$ dimension. The selection function $s_w(x)$ is then defined as

$$s_w(x) = \sum_{i=1}^{n} w_i \overline{f}_i(x) \tag{6}$$

where $w = (w_1, ..., w_n) \in W$ is a weight vector. s_w can be considered to be a simple model of the final choice process where w represent the relative importance of the optimization criteria and the solution selected is the one minimizing s_w. The solution set quality measure q is then defined as

$$q(X) = E(min\{s_w(x) \mid x \in X\}) \tag{7}$$

where the expected value E is computed over all $w \in W$, a given weight space. Varying w over W generates a class of selection functions s_w, thus accounting for a range of different user preferences. For example, W can be defined by assuming that each w_i is uniformly distributed on $[0,1]$ and that the w_is are independent.

For practical computation, $q(X)$ can be estimated by

$$\hat{q}(X) = \frac{1}{N} \sum_{k=1}^{N} min\{s_{w(k)}(x) \mid x \in X\} \tag{8}$$

where each $w(k)$ is generated according to the definition of W and N is the sample size.

The previous problem concerning the combination of distinct quality indicators is now avoided since (7) defines solution set quality directly as a single-valued quantity. Furthermore, (7) is an implicit definition of set quality in the sense that minimizing (7) still encourages optimization of each of the quality indicators of Section 3.1, but without explicitly stating such quality characteristics.

In addition, the following intuitively desirable properties of q are proven in [5]:

1. Only non-dominated solutions of a set X can contribute to $q(X)$. Consequently, adding a dominated (i.e. poor) solution to a set can not improve its quality.
2. Adding a non-dominated (i.e. good) solution to X either leaves $q(X)$ unchanged or improves $q(X)$.
3. If a set X is clearly better than Y in the sense that every solution in Y is dominated by some solution in X, then the quality of X is always at least as good as that of Y, i.e., $q(X) \leq q(Y)$. This holds independently of the sizes of X and Y.

The last property assures that algorithm A (e.g. a random walk) can never obtain better solution set qualities than algorithm B (e.g. a genetic algorithm) when the sets generated by B dominates those of A. Hence, the problem encountered in Section 4.5 (Figure 10) is eliminated.

A remaining serious limitation of (7) is that, as reflected in 2 and 3 above, $q(X)$ is not guaranteed to improve when a non-dominated solution is added; it may stay unchanged. Specifically, if a non-dominated solution is non-convex relative to the other non-dominated solutions in X, it can never improve $q(X)$ (see [5] for a more detailed discussion of this phenomenon).

Figure 11 illustrates the set quality measure on eight constructed sets s1 through s8, assuming two optimization dimensions. The corresponding estimated set quality values are given in Table 1. Except for sets s5 and s6, set quality decreases strictly with the set number, as seems intuitively reasonable. However, s5 is not better than s6 because the solution with cost (8,7) is non-convex relative to s5.

Table 1. Estimated set qualities for sets s1 through s8, using (8) with $N = 50,000$.

set	s1	s2	s3	s4	s5	s6	s7	s8
$\hat{q}$	0.0001	0.0762	0.2127	0.2642	0.4130	0.4130	0.5232	0.5253

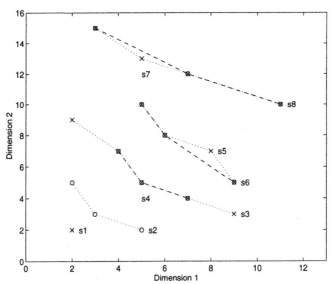

Figure 11. Each set is indicated by either circles connected by a dotted line or crosses connected by a dashed line. s4 is a subset of s3, s6 a subset of s5 and s8 a subset of s7.

5.2 First Experiments

5.2.1 Dynamic Search Performance

The new set quality formulation requires us to modify the dynamic search performance measure. According to the changes in the solution set quality measure outlined in the previous section, we now use the following formula for online performance

$$\frac{1}{t_{final}+1} \sum_{t=0}^{t_{final}} q(X^t). \tag{9}$$

As with the previous dynamic performance measure, minimizing this measure promotes finding a low set quality measure in the shortest time. An issue that will be addressed in future research is that of the relation between the final set quality value and the dynamic performance measure used in our experiments.

5.2.2 Results

The goal of our research is to develop a genetic algorithm that performs well in the context of multiobjective optimization. We performed several IC placement experiments with random walks and previously designed genetic algorithms to evaluate the effectiveness of our proposed dynamic genetic algorithm methods. The results reported in the following sections are obtained from experiments using a real world design consisting of 20 blocks and 248 interconnections[1]. The computational resources for each search was set to 200 cpu seconds on an HP715/64 workstation, which would be much higher in practice.

The random walk operated as describe in Section 4.5. The static genetic algorithm uses static parameter settings for population size, mutation rate, and bias. These values were set according to experience.

The optimized static genetic algorithm used a different set of static parameters for population size, mutation rate, and bias. These values were obtained by optimizing the dynamic search performance measure presented at the beginning of this experimental section. The meta-level optimization was set up as in Section 3.2. The meta-level genetic algorithm used the same parameter values as the meta-level genetic algorithm in Section 4.4. Each of the experiments were allowed to run for 500 generations.

We changed the structure of the controller used in the dynamic genetic algorithm to accommodate the changes in the indicators. Inputs to the new fuzzy controller were: set quality, set quality change since the previous control action, and time left (See Figure 12).

The change in set quality since the previous control action, is measured as follows:

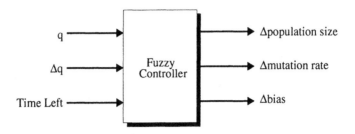

Figure 12. Inputs and outputs of the fuzzy controller used to modulate genetic algorithm parameters.

$$\Delta q(X^t, X^{t-1}) = \frac{q(X^t) - q(X^{t-1})}{q(X^{t-1})}. \tag{10}$$

We added the time input after considering how tasks are usually defined. It is usually the case in problem solving that a deadline is given at the beginning. However, humans will usually alter their strategy by not how much time has elapsed, but by how much time is left. In this new controller, we take the same approach by feeding into it an estimate of how much time is left.

Based on this controller structure, we then tried to find a set of rules, and initial conditions for the outputs, that optimizes the performance measure. The internal structure of the controller used the shared-triangular-membership function representation presented in Section 3.2.1. There was a maximum of 10 rules per output variable. The meta-level genetic algorithm had static parameter settings identical to the meta-level genetic algorithm used to optimize the static genetic algorithm.

The results of the random walk, static genetic algorithm, optimized static genetic algorithm, and dynamic genetic algorithm are given in Table 2 below. Each of the entries were compiled from ten independent runs. The first and second columns report the mean and standard deviations of the various algorithms according to (9). The third and fourth columns report the same statistics on the final output set, $X^{t_{final}}$, obtained by the algorithms.

Table 2. Dynamic and Final q Performance Measures

Algorithm	$\overline{DSM}$	DSM_σ	$\hat{q}(X^{t_{final}})$	$q_\sigma(X^{t_{final}})$
Random Walk	0.406	0.036	0.376	0.031
Genetic Algorithm	0.311	0.029	0.273	0.030
Optimized Genetic Algorithm	0.297	0.015	0.264	0.012
Dynamic Genetic Algorithm	0.299	0.023	0.261	0.017

According to these results, using a genetic algorithm for IC placement improves both the final value and offline measure relative to random search. The optimized static genetic algorithm showed improvements in both the mean final value and mean offline measures over the genetic algorithm with manually set parameter values. One of the most striking results is that the optimized genetic algorithm had much lower standard deviations than the other algorithms.

The dynamic genetic algorithm is very similar to that of the static genetic algorithm. Possible reasons for this result are discussed in Section 6.

6 Summary, Conclusions, and Extensions

The development of good performance indicators is an essential step toward improving the performance of multiobjective search algorithms and understanding them. One approach to characterizing the performance of these algorithms, whose output are solutions sets, is to measure the quality of the solution sets they produce. In this paper, we developed two solution set quality measures that, although based on fundamentally different principles, reward the same set characteristics. These set quality measures were then used to determine parameter settings and parameter control strategies for a genetic algorithm for IC placement.

While the intuition behind the second quality measure presented may seem more opaque than the first measure presented, it better captures our subjective notion of a high quality set. However, the measure does not handle non-convex solution sets appropriately and introduces some amount of sampling noise.

The offline measures derived from the set quality measures and optimized at the meta-level are also subject to noise due to 1) the limited number of runs of the algorithm evaluated and 2) the limited number of circuits considered by the algorithm evaluated. At this point, the impact of these noise factors is not clear.

Results comparing genetic algorithms using empirically determined parameter settings and genetic algorithms with optimized parameter settings show that performance can be significantly improved by the meta-level optimization approach presented in this paper. Because these results hold for both set quality measures, the proposed approach to design high performance algorithms is feasible. The key points for making the approach practical, however, are 1) formulating the appropriate search performance measure and 2) dealing with various forms of noise inherent to meta-level optimization.

Using the second set quality measure, the dynamic parameter control strategy does not improve over an optimized static parameter strategy. In addition to the noise sources identified above, another possible reason for this result is the increased search space size. Two approaches that address this issue are to 1) give the meta-level algorithm more time to search, or 2) find more compact representations for the fuzzy controller.

7 Acknowledgments

This research is supported in part by NASA Grant NCC-2-275, ONR Grant N00014-96-1-0556, and BISC program of UC Berkeley. The authors would also like to thank Prof. David Wessel and the Center for New Music and Audio Technologies at UC Berkeley for use of computing resources. Dr. Esbensen is supported mainly by the Danish Technical Research Council.

8 References and Related Publications

[1]   Asanovic, A., Beck, J., "T0 Engineering Data, Revision 0.14," Technical Report, International Computer Science Institute, Berkeley, CA, 1994.

[2]   Cohoon, J.P., Hedge, S.U., Martin, W.N., Richards, D., "Distributed Genetic Algorithms for the Floorplan Design Problem," *IEEE Transactions on Computer-Aided Design*, Vol. 10, pp. 484-492, April 1991.

[3]   DeJong, K.A. (1975) An Analysis of the Behavior of a Class of Genetic Adaptive Systems, Ph.D. Dissertation, University of Michigan, University Microfilms No. 68-7556.

[4]   Esbensen, H., Kuh, E.S., "An MCM/IC Timing-Driven Placement Algorithm Featuring Explicit Design Space Exploration," *Proc. of the 1996 IEEE Multi-Chip Module Conference*, pp. 170-175, 1996.

[5]   Esbensen H., Kuh, E.S., ``Design Space Exploration Using the Genetic Algorithm," *Proc. of the IEEE International Symposium on Circuits and Systems*, 1996 (to appear).

[6]   Fonseca, C.M. and Fleming, P.J. (1993) "Genetic Algorithms for Multiobjective Optimization: Formulation, Discussion, and Generalization," in *Proc. of the Fifth International Conference on Genetic Algorithms*, ed. S. Forest, San Mateo, CA: Morgan Kaufmann.

[7]   Grefenstette, J.J. (1986) "Optimization of Control Parameters for Genetic Algorithms", *IEEE Trans. on Systems, Man, and Cybernetics*, Vol. 16, No. 1.

[8]   Holland, J. H. (1975) *Adaptation in Natural and Artificial Systems*, MIT Press, Cambridge, MA.

[9]   Lee, M. A. (1995) "On Genetic Representation of High Dimensional Fuzzy Systems," *Proc. of NAFIPS'95*, College Park, MD, IEEE Computer Science Press, pp. 752-757.

[10]   Lee, M. A. and Takagi, H. (1993) "Integrating Design Stages of Fuzzy Systems using Genetic Algorithms," *Proc. IEEE Int. Conf. on Fuzzy Systems* (FUZZ-IEEE '93), San Francisco, CA, pp.612-617.

[11]   Lee, M.A. and Takagi, H. (1993) "Dynamic Control of Genetic Algorithms using Fuzzy Logic Techniques," in *Proc. of the Fifth International Conference on Genetic Algorithms*, ed. S. Forest, San Mateo, CA: Morgan Kaufmann.

[12]   Takagi, T. and Sugeno, M. (1985) "Fuzzy Identification of Systems and Its Applications to Modelling and Control," *IEEE Transactions on Systems, Man, and Cybernetics*, Vol.5, No.3, pp. 116-132.

[13]   Whitley, D., "The Genitor Algorithm and Selection Pressure: Why Rank-Based Allocation of Reproductive Trials is Best," *Proc. of the Third International Conference on Genetic Algorithms*, pp. 116-121, 1989.

[14]   Wong, D.F., Liu, C.L., "A new algorithm for floorplan design," *Proc. of the 23rd Design Automation Conference*, pp. 101-107, 1986.

Structure Identification of Acquired Knowledge in Fuzzy Inference by Genetic Algorithms

Shohachiro Nakanishi†, Akihiro Ohtake†, Ronald R. Yager‡,
Shinobu Ohtani†, and Hiroaki Kikuchi†

†: Dept. of Electrical Engineering
TOKAI University
1117, Kitakaname, Hiratsuka, Kanagawa, 259-12, JAPAN
Tel: +81-463-58-1211, Ext. 4025
Fax: +81-463-59-4014
‡: Machine Intelligence Institute
Iona College
New Rochell, NY, 10801, USA
Tel: +1-212-249-2047
Fax: +1-212-249-1689

Abstract: In the fuzzy modelling and construction of fuzzy inference rules for fuzzy controllers, it is a very important problem to acquire automatically the knowledge for the objects from only their given data. Many methods for knowledge acquisition have been reported and published in the many journals and proceedings in the conference. However, there are no method to identify the structure of acquired knowledge. Then the authors propose a method to identify the structure of the acquired knowledge for objective systems in the form of the multi-stage fuzzy inference from only their given input and output data by a genetic algorithm

1. INTRODUCTION

A fuzzy inference is the powerful tool as a modelling method for complex and imprecise systems, and fruitful results for them have been reported in the various journals and proceedings in the world. The fuzzy inference is also a very suitable method to represent the knowledge of human experience. However, the fuzzy inference does not possess essentially the learning mechanisms or algorithms in it. Then, in the early stage of the development for the fuzzy modelling and the construction of fuzzy inference, the experts of fuzzy logic, who have a deep knowledge for the objective systems, had to construct the fuzzy inference rules. Recently, it have been reported the various methods to construct automatically the fuzzy inference rules from only given input and output data of the object systems by the learning algorithm such as the neural networks and genetic algorithms [1,2]. From a viewpoint of the knowledge engineering, this fact means that we can acquire the knowledge for the objective systems in the form of fuzzy inference rules from the given their data. This is a important and great progression in the field of fuzzy inference and knowledge acquisition. However, in the automatic extraction of the knowledge for the objective systems from the given data, it has been assumed that the structure of the fuzzy inference has been a single stage multi-fold fuzzy inference.

In the complex real systems, there are many systems in which the input information (input data) is transformed in many times, and finally transformed to the output information (output data). This fact means that the systems include many structural steps transforming internal information. That is, the systems are composed of the hierarchical structure in information flow. In the knowledge engineering, it is desired that the acquired knowledge for the systems represents their feature and structure at the same time. The structural representation of acquired knowledge is a natural requirement when the objective systems have hierarchical structures of the information flow.

On the other hand, a genetic algorithm (GA) developed by John Holland [4] is a methodology of a probabilistic search, learning, and optimization based on the mechanics of natural selection and natural genetics. The genetic algorithm is mostly applied to the real systems as the problem solving method of combinatorial optimization problems. The genetic algorithm is also an elegant methodology of the machine learning. In the applications of GA to fuzzy systems, it has been reported that we could use GA as the learning algorithm of self organizing methods formulating the fuzzy rules from only given input and output data of the systems. In those application, T. Fukuda and et al. [3] proposed an elegant method to reduce the number of the fuzzy rules in the form of the hierarchical structure from the given data by genetic algorithm. The main purpose of the paper [3] is to construct a new type system of self-tuning fuzzy inference in which the number of fuzzy rules is reduced to smaller number by using the hierarchical fuzzy inference structure. In the paper [3], an initial form of the hierarchical structure like a binary decision tree diagram in the knowledge engineering is assumed to simplify their problems in the application of GA. This assumption is a very good idea for their purpose. However, in the structure identification in the form of a multistage fuzzy inference, we have to determine the system structure and the combinations of the input variables at a same time. For this purpose, it needs no restriction for the hierarchical structure and grouping the input variables in the fuzzy inference. Especially, [3] cannot determine all combinations of the input variables and the structure under the assumption like a binary decision tree form.

The authors propose a method to identify the knowledge structure of the objective systems in the form of the fuzzy inference rules from only their input and output data by genetic algorithms. In the method, we construct the fuzzy inference rules to identify the objective system structure and grouping of the input variables by GA. The acquired fuzzy rules represent the knowledge for the objective systems, and are constructed in the form of multi-stage fuzzy inference rules of hierarchical style. If we can represent the acquired knowledge in the hierarchical structure form which is formulated by fuzzy inference rules, we can easily understand the intelligent structure of the objective systems. Finally, we apply the method to various kind of systems which are represented by many nonlinear functions, and identify their systems by the method proposed in this paper.

2. FUZZY INFERENCE

The fuzzy inference is usually represented by a set of the fuzzy inference rules as follows:

If x_1 is A_{i1} and x_2 is A_{i2} and .... and x_n is A_{in} then y is B_i,   (1)

$$(i = 1, 2, ... , m),$$

where A_{i1} and B_i are the fuzzy sets characterized by the membership function, and x_i and y stand for the input and output variables of the fuzzy inference, respectively. In general, the fuzzy inference is represented by several input variables and many inference rules as shown in Eq. (1), and the fuzzy inference of this style is called a **multi-fold fuzzy inference** as shown in Fig.1. This styled fuzzy inference is generally used in most applications such as a fuzzy control, expert systems, and other systems. As shown in Fig.1, the ordinary fuzzy inference is composed of a single stage inference structure.

From theoretical point of view, the fuzzy inference allows to use "n" (a large number) input variables. However, in the real applications, the number of the input variables are

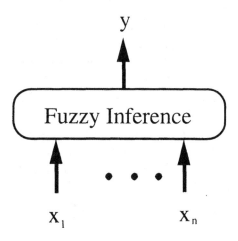

Fig.1 Multi-fold fuzzy inference

experientially limited in a several number of inputs, because the fuzziness of the output (inference result) of the multi-fold fuzzy inference become to extremely increase. In the most real complex systems with a large number of variables, there is a lot of cases which we can easily decrease these variables to a small number of classes from the system's structure, and make several groups of input variables for fuzzy inference. This means that we can make several groups of the fuzzy inference rules. In such a case, we can use a **multi-stage fuzzy inference** as shown in Fig.2. As shown in Fig.2, the outputs

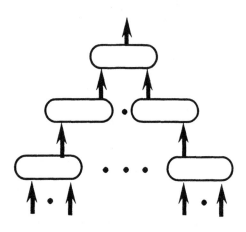

Fig.2 Multistage fuzzy inference

(inference results) in lower stage, in multi-stage fuzzy inference, became to the inputs to the inference in higher stage. That is, we can easily understand the structure of the acquired knowledge from the hierarchical inference, if we can acquire and represent the knowledge for the objects in the form of the multi-stage fuzzy inference style. Then, the multi-stage fuzzy inference is a powerful method to identify the system structure in the modeling of the complex systems.

3. STRUCTURE IDENTIFICATION OF ACQUIRED KNOWLEDGE

It is a well known fact that the fuzzy inference is a good methodology to the modelling and identification of the complex systems containing many imprecise factors, and the fuzzy modelling are widely used in various fields. In the early stage in the development of fuzzy modelling, the fuzzy inference rules are formulated by the experts who have a deep knowledge for the objective systems. Recently, some researchers reported the elegant methods to automatically formulate the fuzzy inference rules identifying the objective systems from only input and output data of the systems by the learning methods of neural networks or genetic algorithms. This fuzzy modelling means to acquire the knowledge for the objective systems in the sense of knowledge engineering, and acquired knowledge is represented in the style of fuzzy inference rules. However, the style of constructed fuzzy inference is only a single stage structure as shown in Fig.1, because the single stage form is assumed in those methods.

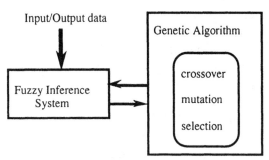

Input/Output data

Fig.3 Structure identification System of acquired knowledge

In the most real complex systems, there are many systems which have own structures naturally. Then, the fuzzy modelling, i.e. the acquired knowledge should reflect the system structure. Then, we propose a method to identify the system structure in the form of multi-stage fuzzy inference from only given system's input and output data by the genetic algorithm as shown in Fig.3. That is, the main purpose of this paper is to construct the fuzzy inference system (acquired knowledge) whose structure is the same structure for the objective system. For this purpose, we have to solve two problems at the same time;

(1) the structural identification of acquired knowledge,

(2) combinatorial identification of the system's input variables under the structure determined in (1).

After solving these problems, we have to construct the optimal fuzzy inference under the conditions of the structure and the combination of the input variables.

For example to describe the contribution of input variables for the system, let us consider the case of the system with three input variables and a single output. In this case, we can consider two types as the structures of inference such as a single stage fuzzy inference and two stage fuzzy inference (see Fig.4). In the two-stage fuzzy inference, the only two variables become to the input variables in the first stage inference, and the output of the first stage inference and a rest input variable become to the input variables in the second stage inference as shown in Fig.4. As shown in Fig.5, there are three combinations of input variables in the first stage

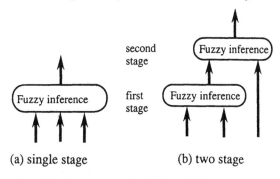

(a) single stage

(b) two stage

Fig.4 An example of fuzzy inference structure

inference under the same structure in this case. This fact shows that there are many combinations of the input variables in the first stage under the same structure.

3.1. Encoding Technique in GA

We have to determine the structure of fuzzy inference and the combination of the input variables at the same time as previously mentioned. We use the genetic algorithm to solve the combinatorial problems of structural identification and combination of input variables in multi-stage fuzzy inference. Then, the chromosome in GA has to include the genes representing the structure

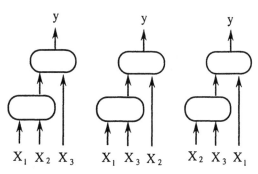

Fig.5 An examole of combination of input variables

information and the combinatorial information of input variables. From this reason, we have to make the chromosomes including information of the structure, combination of input variables, and fuzzy inference rules under the assumed structure. It is noted that each chromosome in the population represents a fuzzy model.

3.1.1. Encoding for inference structure

A chromosome is mainly composed of two parts such as

(1) a structure part, and

(2) a fuzzy inference part,

as shown in Fig. 6, from the reason above mentioned. As shown in Fig.6, the first part of the chromosome represents the structure and

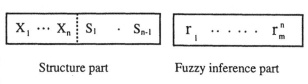

Structure part   Fuzzy inference part

Fig.6 Structure of chromosome

combination of the input variables under that structure, and the second part also shows the fuzzy inference rules under that structure. Fig.7 shows a graph of structural connection, and is the symbolical representation of the multi-stage fuzzy inference. A symbol S_j in Fig. 7 represents a "stage" in multi-stage fuzzy inference. In the case shown in Fig.7, the number of possible maximum stage is n-1, and Fig. 7 shows possible all connections from input variables to each stage in the multi-stage fuzzy inference.

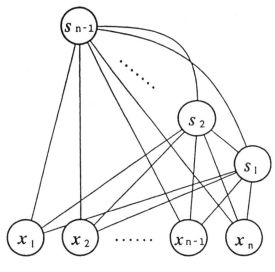

Fig.7. A model of multi-stage fuzzy inference.

Fig. 8 shows the detail of the structure part representing the connections in chromosome. As shown in Fig.8, the structure part is also composed of a input variable part and stage part which represents "stage" in the multi-stage fuzzy inference. The input part is represented by a string of n elements in case of "n" input variables, and the stage part by a string of n-3 elements as shown in Fig. 8. As can be seen in Fig. 7, the n-2 stage S_{n-2} is necessary connected to the final stage S_{n-1} (n-1 stage). Therefore, it does not need to represent the combination of the connection between these units S_{n-2} and S_{n-1} for encoding. Moreover, the information for the single stage structure does not need in the stage part, because the single stage structure happens when all values in the input part take the same value.

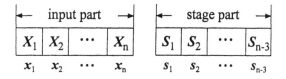

structure part

Fig.8. Detail of structure part

Each part in the chromosome is represented by the string of natural numbers in this paper. Now, we describe the detail of the encoding method for the structure part except the fuzzy inference part, because the encoding method for the fuzzy inference part will be represented in the following article.

(a) **Input variable part:** The input variable part is composed of the string of the natural number from 1 to n-1 as shown in Fig. 9-(a). In this figure, the first position in the input part is corresponding to the input variable "x_1." Each natural number in the locus represents the stage which is connected to the input variable corresponding to that locus. For example, if the value of the variable "x_4" is 2 as shown in Fig.9-(a), the variable x_4 is directly connected to the second stage S_2 in the multi-fuzzy inference system. That is, the natural number in the input part represents the stage which should be connected to the variable .

(b) **Stage part:** The stage part represents the connection between the stages. Let us consider the case of 5 input variables as shown in Fig. 9, as an example. In this case, the stage S_1 is connected to one of the stages from S_2 to S_4. Therefore, the value of S_1 takes the natural number from 2 to 4. The stage S_2 has to be connected to one of the stages from S_3 to S_4, except S_1. Then, S_2 only takes the value 3 or 4. The stage S_3 is only connected to S_4. From this fact, we can generally say that the stage S_{n-2} must be connected to final stage S_{n-1}, then we does not to check the connection for S_{n-2}. From this reason, the stage part is represented by the string of n-3 elements from S_1, S_2, ... , S_{n-3}. It is noted that the range of the values in each locus in the stage part is different as mentioned above.

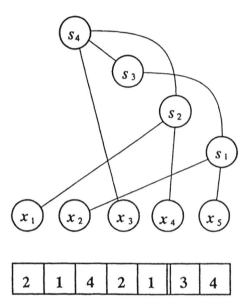

2	1	4	2	1	3	4
x_1	x_2	x_3	x_4	x_5	S_1	S_2

(a) Encoding and connection.

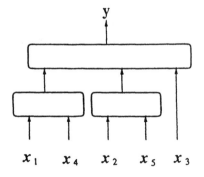

(b) Structure of multi-stage fuzzy inference.

Fig.9. An example of structure part for five variables.

We show an example of encoding for five input variables as shown in Fig.9. In this case, the number of possible stages become to 4 (n - 1 = 5-1). The input variable part of the structure part is composed of five loci, and the stage part is also composed of two (n - 3= 5-3=2) loci in this case. In the input variable part, each natural number in the locus represents the target "stage" connected to the variable. From Fig.9 (a), we can see that the values corresponding to the variables x_1 and x_4 are the same number "2," and then connected to S_2. The variables x_2 and x_5 also connected to S_1. The values for S_1 and S_2 in stage part are 3 and 4, respectively. Then, S_1 connected to S_3, and S_2 is also connected to S_4 as shown in Fig.9-(a). Fig.9-(b) represents the structure of multi-stage fuzzy inference shown by Fig.9-(a). As shown in Fig.9, a stage S_3 is absorbed to S_4 and become to the dummy stage. Then stage S_1 is directly connected to S_4.

If all numbers in the input part represent the same value, the structure of fuzzy inference become to the single stage, that is, multi-fold fuzzy inference. There are some cases in which the encoding of the structure part represents the single stage. Let us consider the coding as shown in Fig. 10, as an example. In this case, S_1 becomes to dummy stage. This fact means that the variable x_5 is connected to the S_2. We also understand that the stages S_3 and S_4 are dummy stages, then the structure represented in this coding is a single stage, that is, multi-fold fuzzy inference.

Both of the input variable part and stage part fully describe the combinations of connections between input variables and stages, and between stages, respectively. It is noted that the input part expresses the structure of the single stage fuzzy inference when all the values in the loci in that part take the same number.

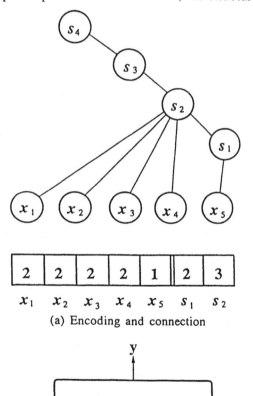

(a) Encoding and connection

(b) Structure represented above encoding

Fig10. An example of single stage

If there exist the different values in the input variable part, the inference structure become to the multi-stage fuzzy inference.

3.1.2. Encoding for fuzzy inference

We have to identify the both of the inference structure and optimum fuzzy inference rules under the structure determined in a previous article at the same time. We have already discussed the encoding for the inference structure identification in the previous article. Now, we explain the encoding for the determination of the optimum fuzzy inference rules under the assumed structure by the genetic algorithm. As the chromosome is composed of the two parts such as a structure part and fuzzy inference part as shown in Fig.6 previously mentioned, we focus our attention to the fuzzy inference part here.

Let us assume the number of membership functions on each input variable, i. e. the number of the fuzzy partition of the input space, be "m," and the number of the membership functions for the consequent part of fuzzy rules be "p." The possible maximum number of the total fuzzy.rules, happen in the case of a single stage fuzzy inference structure, is "m^n," because the number of the input variables is n. The number of the fuzzy rules decreases from the maximum number when the structure becomes to multi-stage inference structure. Then, the number m^n of genes is needed as the size of the fuzzy rule part for representing the single stage type fuzzy inference, which is maximum size.

Each fuzzy rule under the given structure is numbered, and is arranged in the fuzzy inference part from the top of this part in accordance with the numbering order. Then, the position of the locus in the fuzzy inference part represents the ordering number of fuzzy inference rule (see Fig.11). The value of a symbol r_i in the rule part in Fig.6 takes one of the labels of the membership functions in the consequent part of the fuzzy rule.

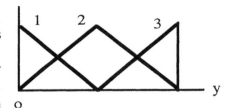

Fig.11. Membership function of consequent part

Then, r_i takes a natural number from 1 to p which is the number of fuzzy partition for the consequent fuzzy variable (see Fig.11).

In the case of multi-stage inference structure, number of the fuzzy inference rules become to decrease from the maximum possible number of rules, arising in case of the single stage multi-fold fuzzy inference. In this case, it needs the smaller number of loci than that of maximum rule case. Then, we use necessary loci in the strings of the inference part from the top of them, and we neglect the rest part in the strings. That is, necessary fuzzy rules are encoded from top of strings according to the inference structure which is represented in the structure part in the chromosome.

Let us consider three input variables as the example of the most simple case of the encoding for fuzzy inference part. In this case, "n" become to three, and we assume that the number of membership functions for the variables in both of antecedent and consequent part is three (m,p=3) as shown in

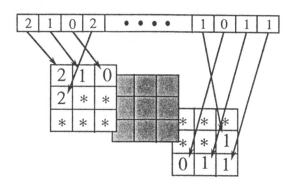

Fig.12. Fuzzy inference part and their correspondence

Fig.11. The number 1, 2, and 3 in Fig.11 represent the label of the membership function. The size of the fuzzy inference part, i.e. size of this string is 3^3 (=27) which is the possible maximum number of fuzzy inference rules in this case.

Fig.12 shows an example of encoding of the fuzzy inference part in the chromosome. The natural number in Fig.12 represents the label of membership function of the consequent part of the fuzzy inference, and a gene corresponds to a fuzzy rule. Each rule is numbered, and is arranged in accordance with the numbering as shown in Fig.12.

4. STRUCTURE IDENTIFICATIONS OF SYSTEMS

The aim of this chapter is to check whether the system proposed in this paper has ability to identify the structure of systems by only their input and output data. For this purpose, we check some systems expressed by linear and nonlinear functions as shown in Table 1. It needs the input and output data as the learning data to identify the systems by our method. For making learning data, we randomly select 100 points in the three dimensional space, and calculate the system outputs by each function listed in Table 1.

Table 1 also shows the simulation results. The second and third columns in Table 1 show the functions representing system behavior and the resultant, system structures from the experiments, respectively. In the third column, a brace $\{\bullet, \bullet\}$ means that the variables in the brace $\{\bullet, \bullet\}$ belong to the same input group, and become to the inputs to the first stage of the inference. The other variables outside of the brace $\{\bullet, \bullet\}$ in the second column and the output of the first stage become to the inputs to the second stage in the inference. Then, Table 1 shows

Table 1. Functions representing system and resultant structure from simulation

No.	Function	Structure
1	$x_1^{x_2+x_3} - 1$	$x_1\{x_2, x_3\}$
2	$\frac{x_1+x_2x_3}{2}$	$x_1\{x_2, x_3\}$
3	$x_1x_2x_3$	$\{x_1, x_2\}x_3, \{x_1, x_3\}x_2$
4	$\frac{2x_1x_2+x_3}{3}$	$\{x_1, x_2\}x_3$
5	$\frac{x_1^4+x_2^3+x_3}{3}$	$\{x_1, x_2\}x_3$
6	$\left|\frac{\sin(x_1\pi)+\cos(x_2x_3\pi)}{2}\right|$	$\{x_1, x_2\}x_3$

Table 2. Factor loadings from factor analysis.

No.	Pattern Factor 1			Pattern Factor 2		
	x_1	x_2	x_3	x_1	x_2	x_3
1	−0.01	**0.77**	**0.71**	**0.91**	−0.27	0.35
2	**0.87**	−0.01	**0.60**	−0.13	**0.93**	0.42
3	−0.01	**0.77**	**0.70**	**0.91**	−0.29	0.35
4	**0.80**	−0.75	−0.05	0.06	0.15	**0.99**
5	0.10	−0.60	**0.86**	**0.89**	0.51	0.17
6	**0.82**	0.02	**0.65**	0.22	**0.92**	−0.34

that all systems represented by the listed functions have a hierarchical structure.

Table 2 shows the results of factor analysis for the same data of the same systems represented in Table 1. The second and third columns in Table 2 represent the factor loadings for a factor 1 and factor 2, respectively. As can be seen from Table 2, the results in the third column show the complementary results of the second column in this case, then we only check the results of second column for the factor analysis.

Let us consider the systems of No.1 and 5 listed in Table 1 and 2. For No.1 system, Table 2 shows high values of factor loadings for x_2 and x_3, on the contrary, and shows extremely low value for x_1. Table 1 also shows that the variables x2 and x3 in No.1 system belong to the same group. This fact means that the calculating operation for x_2 and x_3 is firstly carried out, and total output is calculated finally, in multi-stage fuzzy inference. We can intuitively understand that both results in Table 1 and 2 are valid results from the form of the equation representing No.1 system. For No.5 system, both results in Table 1 and 2 show the same meaning. That is, both results of the simulation and factor analysis for No.1 and No.5 systems show the same structure and the same input combination to the multi-stage fuzzy inference for these systems. Although the results of No.1 and No.5 systems in the Table 2 represent the same content in Table 1, the other results in Table 2 is different from that in Table 1.

Let us consider the No.2 system as an example of the other cases. From the simulation results in Table 1, the variables x_2 and x_3 belong to the same group, and form the first stage in multi-stage fuzzy inference. On the contrary, the results from factor analysis in Table 2 show that x_1 and x_3 should belong to the same group. If we observe the functions representing No.2 system as

$$f(x_1, x_2, x_3) = (x_1 + x_2 x_3)/2,$$

we can intuitively understand that x_2 and x_3 belong to the same calculating group. Then, we can understand that the structure of the inference and grouping of the input variables from the simulation in Table 1 show the more reasonable results than the results of factor analysis, for human sense.

Let us consider No.3 system one more example. The simulation result for No.3 system in Table 1 shows two combinations of the input variables as the equivalent results. On the other hand, the result from factor analysis in Table 2 shows only one result of combination for input variables. However, if we observe the function representing No.3 system as

Table 3. Functions representing system and resultant structure from simulation

FUNCTION	STRUCTURE
$x_1 x_2 x_3$	$\{x_1, x_2\} x_3$
$\|x_1^2 x_2^2 - x_3\|$	$\{x_1, x_2\} x_3$
$\frac{x_1 + x_2 + x_3}{3}$	$\{x_1, x_2, x_3\}$
$\frac{x_1^4 + x_2^3 - x_3}{2}$	$\{x_1, x_2\} x_3$
$\|\sin(x_1^4 + x_2^3 - x_3)\|$	$x_1 \{x_2, x_3\}$
$\frac{x_2 \sin(x_2 x_1)}{x_2^2 + x_3^2}$	$x_1 \{x_2, x_3\}$
$\frac{\sin x_1 \cos x_2 + \sin x_2 \cos x_3 + \sin x_3 \cos x_1}{3}$	$\{x_1, x_3, x_2\}$
$\|\log(x_2 x_3 + x_1) + 0.1\|$	$\{x_1, x_2\} x_3$
$\|x_1 - x_1 \exp \frac{-(x_1 + x_2)}{(x_2 + x_3)}\|$	$x_1 \{x_2, x_3\}$
$\|x_2 - \exp \frac{-x_3 + x_1}{x_3 x_1}\|$	$x_1 \{x_2, x_3\}$
$\frac{\sin x_1 + \cos x_2 + \log x_3}{3}$	$\{x_1, x_2\} x_3$
$\frac{\sin x_1 \cos x_2 \exp x_3 - 1}{3}$	$\{x_1, x_2, x_3\}$
$\|\cos(\sin x_1 + \sin x_2) + x_3\|$	$\{x_1, x_2, x_3\}$
$\sin(3 \sin(\cos(x_1 \sin x_2) + x_3))$	$x_1 \{x_2, x_3\}$
$\sin(\exp(\frac{x_3 x_1}{x_2 + x_1}))$	$\{x_1, x_2\} x_3$
$\|\frac{\sin x_2 \exp(x_3 + x_1)}{\pi}\|$	$\{x_1, x_3\} x_2$

$$f(x_1, x_2, x_3) = x_1 x_2 x_3,$$

we can understand that each variable x_1, x_2, and x_3 equally contribute to the system. Then, we can firstly consider the only one stage multi-fold fuzzy inference. However, the results in Table 1 mean that the summation operation in the function is represented in the same stage, and the product operation in the function is expressed by the series operation in multi-stage inference. That is, summation operation is treated in parallel operation of the multi-fold inference, and product operation is represented by structural operation, i.e. multi-stage. As No.3 system is represented by product operation of three variables, it is natural that the inference is represented by two stage fuzzy inference. In this case, it is considered that each combination of two variables in three input variables is an equivalent. Then, the result from the simulation in Table 1 is more adequate result than that from factor analysis in Table 2.

Table 3 shows the identification results of the other examples. In Table 3, No. 3 system is represented by a single stage inference structure. On the other hand, No. 1 system is represented by two stage inference structure. From over view of these results, we can understand that the product operation in the function representing the system behavior is expressed by a series operation, i.e. hierarchical structure in the inference. On the other hand, summation operation is treated parallel. That is, the summation is generally expressed by a single stage inference. These facts imply that the behavior of the systems represented by the mixed operation of the product and summation is generally expressed by the hierarchical structure of the fuzzy inference.

5. CONCLUSION

We have proposed the method to identify the knowledge structure of the objective systems from only their input and output data by the genetic algorithm. We have shown that the method which is proposed in this paper adequately extracted the structure of knowledge from input and output data by simulations.

References

[1] H.Ichihashi and T. Watanabe: "Learning Control by Fuzzy Models Using a Simplified Fuzzy Reasoning," Journal of Japan Society for Fuzzy Theory and Systems (SOFT), Vol.2, No.3, pp.429/437, 1990. (in Japanese)

[2] Michael A. Lee and H. Takagi: "Integrating Design Stages of Fuzzy Systems using Genetic Algorithms," Proceedings of Second IEEE International Conference on Fuzzy Systems, Vol.1, pp.612/617, 19993.

[3] T. Fukuda, Y. Hasegawa, and K. Shimijima: "Structure Organization of Hierarchical Fuzzy Model using by Genetic Algorithm," Proceedings of FUZZ-IEEE/IFES'95, Vol.1, pp. 295-299, 1995.

[4] J. H. Holland: "Adaptation in Natural and Artificial Systems," Ann Arbor, University of Michigan Press, 1975.

A Fuzzy Classifier System That Generates Linguistic Rules for Pattern Classification Problems

Hisao Ishibuchi, Tomoharu Nakashima and Tadahiko Murata

Department of Industrial Engineering, Osaka Prefecture University
Gakuen-cho 1-1, Sakai, Osaka 593, Japan; e-mail: hisaoi@ie.osakafu-u.ac.jp

Abstract. In this paper, we propose a fuzzy classifier system that can automatically generate linguistic rules from numerical data (*i.e.*, from training patterns) for multi-dimensional pattern classification problems. Classifiers in our approach are linguistic rules such as "If x_1 is *small* and x_2 is *large* and x_3 is *medium* then Class 2 with $CF = 0.9$" where CF is the grade of certainty of this rule. The grade of certainty of each linguistic rule is adjusted in each population by a reward and punishment scheme. A fitness value is also assigned to each linguistic rule, which is determined by its classification performance for training patterns. Our approach is illustrated by computer simulations on two-dimensional pattern classification problems. Learning ability for training patterns and generalization ability for test patterns of our approach are examined by several real-world pattern classification problems involving many features (*i.e.*, many attributes).

1. Introduction

Fuzzy systems based on fuzzy if-then rules have been successfully applied to various control problems [12,16]. In many application tasks, fuzzy if-then rules were usually derived from human experts as linguistic knowledge. Because it is not always easy to obtain fuzzy if-then rules from human experts, recently several methods have been proposed for automatically extracting fuzzy if-then rules from numerical data (for example, see [17,18,21]). Genetic algorithms [4,5] have also been employed for generating fuzzy if-then rules and adjusting the membership functions of fuzzy sets (for example, see [1,10,11,14,19,20]). A fuzzy if-then rule was coded as an individual in Bonarini [1] and Valenzuela-Rendon [20], while a set of fuzzy if-then rules was treated as an individual in Karr [10], Karr & Gentry [11], Nomura *et al.*[14] and Thrift [19].

In the above mentioned studies, fuzzy systems based on fuzzy if-then rules have been mainly applied to control problems. Usually those control problems do not involve many inputs. Thus fuzzy systems can often be represented by fuzzy rule tables. An example of a fuzzy rule table with two inputs (x_1 and x_2) is shown in Fig.1. One advantage of a fuzzy system based on such a fuzzy rule table is its clarity. That is, human users can easily understand each fuzzy if-then rule in the fuzzy rule table because its antecedent and consequent parts are represented by linguistic values such as "*small*", "*medium*" and "*large*".

$$x_1$$

		S	MS	M	ML	L
	S	S	S	S	S	S
	MS	S	S	MS	MS	MS
x_2	M	S	MS	MS	M	M
	ML	S	MS	M	M	ML
	L	S	MS	M	ML	L

Fig.1 An example of a fuzzy rule table (S: *small*, MS: *medium small*, M: *medium*, ML: *medium large*, L: *large*)

A few approaches have been proposed for automatically generating fuzzy if-then rules for pattern classification problems. Ishibuchi *et al.*[6] proposed a simple method for automatically extracting fuzzy classification rules from training patterns. Their method consists of two phases: fuzzy partition of a pattern space into fuzzy subspaces and determination of a fuzzy if-then rule for each fuzzy subspace. A fuzzy partition by a simple fuzzy grid (see Fig.2) was employed in [6]. Because the performance of a fuzzy system strongly depends on the choice of a fuzzy partition, Ishibuchi *et al.*[6] also proposed an idea to simultaneously employ multiple fuzzy partitions in a single fuzzy system. Fig.3 illustrates a single fuzzy system based on five fuzzy partitions where 90 ($= 2^2 + 3^2 + 4^2 + 5^2 + 6^2$) fuzzy if-then rules are simultaneously employed in fuzzy reasoning. The main drawback of this approach is that the number of fuzzy if-then rules becomes enormous. In order to select a small number of significant fuzzy if-then rules by removing unnecessary rules, a genetic-algorithm-based approach was proposed by Ishibuchi *et al.*[7]. In their approach, first a large number of candidate fuzzy if-then rules (*e.g.*, 90 rules in Fig.3) were generated from training patterns, then a small number of significant rules were selected by a genetic algorithm to construct a compact fuzzy system with high classification performance. The rule selection problem in [7] can be viewed as a kind of a knapsack problem. In [7], a subset S of candidate fuzzy if-then rules was represented by a string: $S = s_1 s_2 ... s_N$ where N is the total number of the candidate rules and $s_j = 1$ means that the j-th rule is included in S (*i.e.*, the j-th rule is selected). This approach to the rule selection was extended to the case of rectangular fuzzy if-then rules in [8]. Examples of fuzzy partitions for generating rectangular fuzzy if-then rules are shown in Fig.4 where 36 fuzzy subspaces (*i.e.*, 36 fuzzy if-then rules) are generated by nine fuzzy partitions. As we can see from the coding $S = s_1 s_2 ... s_N$ of a rule set in the genetic-algorithm-based rule selection methods [7,8], the string length (*i.e.*, N) is equal to the total number of candidate fuzzy if-then rules. While these methods work well for low-dimensional pattern classification problems involving a few features (*i.e.*, a few attributes), they can not be applied to high-dimensional pattern classification problems involving many features because the string length becomes intractably large. For example, even if we use only six fuzzy sets for each axis of a pattern space as shown in Fig.4, the total number of the candidate fuzzy if-then rules is more than one million for an eight-dimensional pattern classification problem (*i.e.*, $N = 6^8 > 1,000,000$).

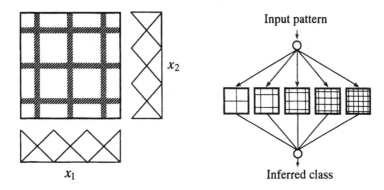

| **Fig.2** An example of a fuzzy partition by a simple fuzzy grid | **Fig.3** A single fuzzy system with five fuzzy partitions |

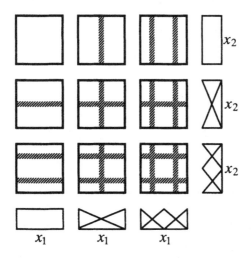

Fig.4 Nine fuzzy partitions for rectangular fuzzy if-then rules

In this paper, we propose a fuzzy classifier system for constructing a compact fuzzy system based on a small number of linguistic classification rules. In our fuzzy classifier system, antecedent fuzzy sets of fuzzy if-then rules are restricted within linguistic values. An example of a set of linguistic values is shown in Fig.5 where the membership functions of six linguistic values are depicted. The choice of linguistic values used for antecedent fuzzy sets depends on each pattern classification problem and should be done by human experts. In computer simulations in this paper, we use the six linguistic values in Fig.5. The main advantage of our fuzzy classifier system is its applicability to multi-dimensional pattern classification problems involving many features. In our approach, each linguistic classification rule is treated as an individual (*i.e.*, as a classifier), and each population consists of a fixed number of linguistic classification rules (*e.g.*, 100 rules). Therefore neither the string length nor the population size increases exponentially when the dimension of pattern classification problems increases. This leads to the applicability of our approach to multi-dimensional pattern classification problems with many features.

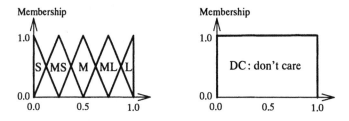

Fig.5 Membership functions of six linguistic values (S: *small*, MS: *medium small*, M: *medium*, ML: *medium large*, L: *large*, DC: *don't care*)

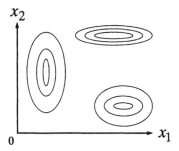

Fig.6 Examples of fuzzy if-then rules in a network architecture

The outline of our fuzzy classifier system can be written as follows:

1. Generate an initial population of linguistic classification rules.
2. Train each rule in the current population.
3. Assign a fitness value to each rule in the current population.
4. Generate new linguistic classification rules and replace a part of the current population with the newly generated rules.
5. Iterate 2-4.

One alternative approach to the construction of a compact fuzzy system is to use a network architecture similar to radial basis function networks [9,13]. Examples of fuzzy if-then rules in such a network architecture are illustrated in Fig.6 where a set of nested ellipsoids corresponds to a fuzzy if-then rule. Each ellipsoid corresponds to a contour line of the membership function of the antecedent part of each fuzzy if-then rule. While this approach based on such a network architecture may be applicable to multi-dimensional pattern classification problems, we lose an inherent advantage of fuzzy-rule-based systems (*i.e.*, the clarity of each rule). It is not easy to linguistically understand each fuzzy if-then rule (*i.e.*, each set of nested ellipsoids) in Fig.6.

2. Pattern Classification by Linguistic Classification Rules

2.1 Rule Generation

Let us assume that our pattern classification problem is a c-class problem in an n-dimensional pattern space $[0,1]^n$. When a given pattern classification problem does not satisfy this assumption, all attribute values are normalized into real numbers in the unit interval $[0,1]$ as a pre-processing procedure for our approach. In Section 4, we use this pre-processing procedure in the application of our approach to real-world pattern classification problems. We also assume that m training patterns $x_p = (x_{p1}, x_{p2}, ..., x_{pn})$, $p = 1,2,...,m$ are given from c classes ($c \ll m$). Because the pattern space is $[0,1]^n$, $x_{pi} \in [0,1]$ for $p = 1,2,...,m$ and $i = 1,2,...,n$.

In this paper, we use the following linguistic classification rules for our pattern classification problem:

Rule R_j : If x_1 is A_{j1} and ... and x_n is A_{jn} then Class C_j with $CF = CF_j$,

$$j = 1,2,...,N, \quad (1)$$

where R_j is the label of the j-th fuzzy if-then rule, $A_{j1},...,A_{jn}$ are linguistic values such as "*small*", "*medium*" and "*large*" (see Fig.5), C_j is the consequent class (*i.e.*, one of the given c classes), CF_j is the grade of certainty of the linguistic classification rule R_j, and N is the total number of linguistic classification rules. Our linguistic classification rules in (1) are different in the following points from fuzzy if-then rules usually used in control problems:

(i) The consequent of each of our linguistic classification rules in (1) is one of the given classes (*e.g.*, Class 1) while it is a linguistic label (*e.g.*, small) in the conventional framework of fuzzy reasoning.

(ii) The grade of certainty is assigned to each of our linguistic classification rules while each rule is evenly handled in the conventional framework of fuzzy reasoning.

It should also be noted that the grade of certainty CF_j is different from the fitness value of each rule. The fitness value is used in a selection operation of our fuzzy classifier system, while CF_j is used in fuzzy reasoning for classifying new patterns.

Because the pattern space is the n-dimensional hypercube $[0,1]^n$, each linguistic value A_{ji} is a fuzzy number on the unit interval $[0,1]$ as shown in Fig.5. Let us assume that each axis of the n-dimensional pattern space has K linguistic values. In this case, the total number of linguistic classification rules is K^n (*i.e.*, $N = K^n$). The choice of linguistic values for each axis of the pattern space depends on characteristic features of the pattern classification problem at hand, and should be done by human experts. In computer simulations in this paper, we use the six linguistic values in Fig.5 for each axis of the *n*-dimensional pattern space. Thus the total number of linguistic classification rules in the computer simulations is 6^n (*i.e.*, $N = 6^n$). For example, we can generate the following 36 linguistic classification rules for a two-dimensional pattern classification problem:

R_1: If x_1 is *small* and x_2 is *small* then Class C_1 with $CF = CF_1$,
R_2: If x_1 is *small* and x_2 is *medium small* then Class C_2 with $CF = CF_2$,
...   .
R_{36}: If x_1 is *don't care* and x_2 is *don't care* then Class C_{36} with $CF = CF_{36}$.

Fuzzy subspaces corresponding to these 36 linguistic classification rules are shown in Fig.7.

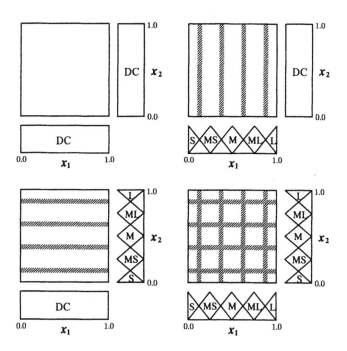

Fig.7 Fuzzy subspaces corresponding to the 36 linguistic classification rules for a two-dimensional pattern classification problem

Each linguistic classification rule R_j in (1) can be generated from the training patterns by the following procedure [6]:

[Rule Generation Procedure]

Step 1: Calculate $\beta_{\text{Class }h}(R_j)$ for Class h ($h = 1,2,...,c$) as

$$\beta_{\text{Class }h}(R_j) = \sum_{x_p \in \text{Class }h} \mu_{j1}(x_{p1}) \cdot ... \cdot \mu_{jn}(x_{pn}), \ h = 1,2,...,c, \tag{2}$$

where $\beta_{\text{Class }h}(R_j)$ is the sum of the compatibility of training patterns in Class h to the linguistic classification rule R_j in (1), and $\mu_{ji}(\cdot)$ is the membership function of the linguistic value A_{ji} in (1).

Step 2: Find Class $\hat{h}_j$ that has the maximum value of $\beta_{\text{Class }h}(R_j)$:

$$\beta_{\text{Class }\hat{h}_j}(R_j) = \text{Max}\{\beta_{\text{Class }1}(R_j), \ ..., \ \beta_{\text{Class }c}(R_j)\}. \tag{3}$$

If two or more classes take the maximum value, the consequent C_j of the rule R_j can not be determined uniquely. In this case, let C_j be ϕ. If a single class takes the maximum value, let C_j be Class $\hat{h}_j$.

Step 3: If a single class takes the maximum value of $\beta_{\text{Class }h}(R_j)$, the grade of certainty CF_j is determined as

$$CF_j = (\beta_{\text{Class }\hat{h}_j}(R_j) - \overline{\beta}) / \sum_{h=1}^{c} \beta_{\text{Class }h}(R_j), \tag{4}$$

where $\overline{\beta} = \sum_{h \neq \hat{h}_j} \beta_{\text{Class }h}(R_j) / (c-1)$.

This procedure is used for generating an initial population of linguistic classification rules in our fuzzy classifier system. This procedure is also used for determining the consequent class C_j and the grade of certainty CF_j of each linguistic classification rule generated by genetic operations such as crossover and mutation. The grade of certainty CF_j determined by this procedure is adjusted by a learning algorithm in our fuzzy classifier system.

2.2 Fuzzy Reasoning

By the rule generation procedure in the last subsection, we can generate N linguistic classification rules in (1). Let us denote a subset of these N linguistic classification rules by S. When the rule set S is given, a new pattern x_p is classified by the following procedure in the classification phase [6]:

[Fuzzy Reasoning Procedure]

Step 1: Calculate $\alpha_{\text{Class }h}(x_p)$ for Class h ($h = 1,2,...,c$) as

$$\alpha_{\text{Class }h}(x_p) = \text{Max}\{\mu_{j1}(x_{p1}) \cdot ... \cdot \mu_{jn}(x_{pn}) \cdot CF_j \mid C_j = \text{Class }h \text{ and } R_j \in S\},$$
$$h = 1,2,...,c. \tag{5}$$

Step 2: Find Class h_p^* that has the maximum value of $\alpha_{\text{Class }h}(x_p)$:

$$\alpha_{\text{Class }h_p^*}(x_p) = \text{Max}\{\alpha_{\text{Class }1}(x_p), \, ..., \, \alpha_{\text{Class }c}(x_p)\}. \tag{6}$$

If two or more classes take the maximum value, then the classification of x_p is rejected (*i.e.*, x_p is left as an unclassifiable pattern), else assign x_p to Class h_p^*.

3. Fuzzy Classifier System

3.1 Generation of an Initial Population

Let us denote the number of linguistic classification rules in each population in our fuzzy classifier system by N_{pop} (*i.e.*, N_{pop} is the population size). To construct an initial population, N_{pop} linguistic classification rules are generated by the rule generation procedure in Section 2. Antecedent linguistic values of each linguistic classification rule are randomly selected, and the consequent class C_j and the grade of certainty CF_j of each rule is determined by the rule generation procedure in Section 2.

3.2 Learning of Linguistic Classification Rules

Each of the given training patterns is classified by linguistic classification rules in the current population using the fuzzy reasoning procedure in Section 2. From the fuzzy reasoning procedure, we can see that a training pattern x_p is classified by the linguistic classification rule $R_{\hat{j}}$ that satisfies the following relation:

$$\mu_{\hat{j}1}(x_{p1}) \cdot \, ... \, \cdot \mu_{\hat{j}n}(x_{pn}) \cdot CF_{\hat{j}} = \text{Max}\{\mu_{j1}(x_{p1}) \cdot \, ... \, \cdot \mu_{jn}(x_{pn}) \cdot CF_j \mid R_j \in S\}, \tag{7}$$

where S is the set of the linguistic classification rules in the current population.

When x_p is correctly classified by the linguistic classification rule $R_{\hat{j}}$, the grade of certainty $CF_{\hat{j}}$ of this rule is increased as the reward of the correct classification as follows [15]:

$$CF_{\hat{j}}^{new} = CF_{\hat{j}}^{old} + \eta_1 \cdot (1 - CF_{\hat{j}}^{old}), \tag{8}$$

where η_1 is a positive learning rate for increasing the grade of certainty. On the contrary, when x_p is misclassified by the linguistic classification rule $R_{\hat{j}}$, the grade of certainty $CF_{\hat{j}}$ of this rule is decreased as the punishment of the misclassification as follows [15]:

$$CF_{\hat{j}}^{new} = CF_{\hat{j}}^{old} - \eta_2 \cdot CF_{\hat{j}}^{old}, \tag{9}$$

where η_2 is a positive learning rate for decreasing the grade of certainty.

This reward and punishment scheme is iterated in each generation. Let us denote the number of iterations of this learning scheme by N_{learn}. This means that the learning of the grade of certainty is iterated N_{learn} times for each training pattern (*i.e.*, N_{learn} epochs) in each generation.

3.3 Fitness Evaluation

In order to assign a fitness value to each linguistic classification rule in the current population, all the given training patterns are classified by linguistic classification rules in the current population using the fuzzy reasoning procedure in Section 2. Based on the classification results, the fitness value of each linguistic classification rule R_j is defined as

$$fitness(R_j) = w_{NCP} \cdot NCP(R_j) - w_{NMP} \cdot NMP(R_j), \tag{10}$$

where $NCP(R_j)$ is the number of correctly classified training patterns by R_j, $NMP(R_j)$ is the number of misclassified training patterns by R_j, and w_{NCP} and w_{NMP} are non-negative weights for $NCP(R_j)$ and $NMP(R_j)$, respectively. In (10), w_{NCP} and w_{NMP} can be viewed as the reward for the correct classification and the punishment for the misclassification, respectively. The fitness value of each rule is calculated in every population.

3.4 Selection

A pair of linguistic classification rules are selected from the current population to generate new rules for the next population. Each linguistic classification rule in the current population (*i.e.*, in the rule set S) is selected by the following selection probability:

$$P(R_j) = \frac{fitness(R_j) - fitness_{min}(S)}{\sum_{R_i \in S} \{fitness(R_i) - fitness_{min}(S)\}}, \tag{11}$$

where $fitness_{min}(S)$ is the minimum fitness value of the linguistic classification rules in the current population S. This procedure is iterated until a pre-specified number of pairs of linguistic classification rules are selected.

3.5 Crossover

From each selected pair of linguistic classification rules, two rules are generated by the uniform crossover for the antecedent linguistic values. The uniform crossover for the antecedent linguistic values is illustrated in Fig.8. It should be noted that only the antecedent linguistic values of each selected pair of linguistic classification rules are mated. The consequent class and the grade of certainty of each generated rule are determined after a mutation operation by the rule generation procedure in Section 2.

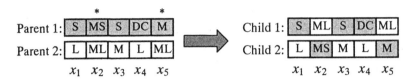

Fig.8 Uniform crossover for the antecedent linguistic values (∗ denotes a crossover position)

3.6 Mutation

Each antecedent linguistic value of the generated linguistic classification rules by the crossover operation is randomly replaced with a different linguistic value. This mutation operation is illustrated in Fig.9. As in the crossover operation, the mutation operation is applied to only the antecedent linguistic values. The consequent class and the grade of certainty of each of the linguistic classification rules generated by the crossover and mutation operations are determined by the rule generation procedure in Section 2.

Fig.9 Mutation for the antecedent linguistic values (∗ denotes a mutation position)

3.7 Replacement

A certain number of linguistic classification rules (say, N_{rep}) in the current population are replaced with newly generated rules by the crossover and mutation operations. In our fuzzy classifier system, all the linguistic classification rules in the current population are arranged in decreasing order of the fitness values, then the last N_{rep} rules are replaced with the newly generated rules. That is, the worst N_{rep} rules with the smallest fitness values are removed from the current population and the newly generated linguistic classification rules are added.

In order to generate N_{rep} linguistic classification rules by the crossover and mutation operations, $N_{rep}/2$ pairs of linguistic classification rules are selected in the selection operation. Then two linguistic classification rules are generated from each pair by the crossover operation.

3.8 Termination Test

We can use various stopping conditions for terminating the execution of our fuzzy classifier system. In computer simulations in this paper, we used the total number of generations as a stopping condition.

3.9 Algorithm

Our fuzzy classifier system can be written as the following algorithm:

Step 0: Generation of an initial population:
Antecedent linguistic values of each linguistic classification rule are randomly specified. The consequent class and the grade of certainty of each rule are determined by the rule generation procedure in Section 2. In this manner, N_{pop} linguistic classification rules are generated.

Step 1: Learning of the grade of certainty:
The grade of certainty of each rule in the current population is tuned by the learning procedure in Subsection 3.2.

Step 2: Evaluation of each rule:
The fitness value of each rule in the current population is determined by (10).

Step 3: Generation of new rules:
According to the selection probability, $N_{rep}/2$ pairs of linguistic classification rules are selected from the current population. By the crossover and mutation operations, antecedent linguistic values of N_{rep} linguistic classification rules are determined from the selected $N_{rep}/2$ pairs. The consequent class and the grade of certainty of each of the newly generated N_{rep} rules are determined by the rule generation procedure in Section 2.

Step 4: Replacement:
The worst N_{rep} linguistic classification rules in the current population are replaced with the newly generated rules.

Step 5: Termination test:
If a pre-specified stopping condition is not satisfied, return to Step 1.

4. Simulation Results

4.1 Simulation Results on Numerical Examples

We applied our fuzzy classifier system to a classification problem in Fig.10. Because the given problem is a two-dimensional pattern classification problem, we use the following linguistic classification rule:

$$R_j : \text{If } x_1 \text{ is } A_{j1} \text{ and } x_2 \text{ is } A_{j2} \text{ then Class } C_j \text{ with } CF_j.$$

As the antecedent fuzzy sets A_{j1} and A_{j2}, we used the six linguistic values in Fig.5. Thus the total number of linguistic classification rules is $6^2 = 36$. Our problem is to select a compact rule set from these 36 linguistic classification rules.

In computer simulations, we specified parameter values of our fuzzy classifier system as follows:

Population size: $N_{pop} = 2 \sim 10$,
Weights in the fitness function: $w_{NCP} = 1$, $w_{NMP} = 5$,
Crossover probability: 1.0,
Mutation probability: 0.1,
Number of replaced rules in each population: $N_{rep} = 1$,
Iterations of the learning procedure in each generation: $N_{learn} = 0$,
Stopping condition: 500 generations.

Because the classification problem in Fig.10 is simple, we implemented our fuzzy classifier system without the learning procedure.

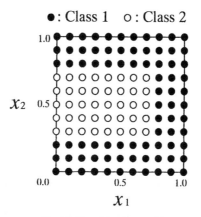

Fig.10 Classification problem

Simulation results are summarized in Table 1. From Table 1, we can see that all the training patterns in Fig.10 can be correctly classified by five linguistic classification rules. The following five rules were generated by our fuzzy classifier system (Attributes with *"don't care"* in the antecedent part are omitted in the following):

If x_1 is *large* then Class 1 with $CF = 1.0$,
If x_2 is *small* then Class 1 with $CF = 1.0$,
If x_2 is *large* then Class 1 with $CF = 1.0$,
If x_1 is *small* and x_2 is *medium* then Class 2 with $CF = 1.0$,
If x_1 is *medium* and x_2 is *medium* then Class 2 with $CF = 1.0$.

From Fig.10, we can see that these linguistic classification rules coincide with our intuitive recognition of the given patterns.

Table 1 Relation between population size and classification rate

Population size	2	3	4	5	6	7	8	9	10
Classification rate (%)	66.9	72.7	97.5	100	100	100	100	100	100

●: Class 1   o: Class 2

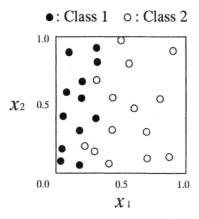

Fig.11 Classification problem

Table 2 Simulation results for the classification problem in Fig.11 with no learning procedure

Population size	2	3	4	5	6	7	8	9	10
Classification rate (%)	73.1	84.6	84.6	84.6	84.6	92.3	92.3	92.3	92.3

Table 3 Simulation results for the classification problem in Fig.11 with the learning procedure

Population size	2	3	4	5	6	7	8	9	10
Classification rate (%)	57.7	73.1	73.1	84.6	84.6	96.2	100	100	100

●: Class 1   o: Class 2

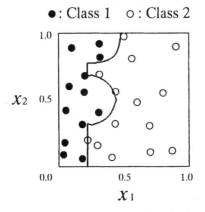

Fig.12 Classification boundary between the two classes by the eight linguistic classification rules obtained by our fuzzy classifier system

We also applied our fuzzy classifier system to the pattern classification problem in Fig.11 in the same manner as in the above computer simulations. Simulation results are summarized in Table 2. From Table 2, we can see that our fuzzy classifier system with no learning procedure could not find a compact rule set with high classification performance.

We also applied our fuzzy classifier system with the learning procedure to the pattern classification problem in Fig.11. The learning procedure with $\eta_1 = 0.001$ and $\eta_2 = 0.1$ in Subsection 3.2 was iterated five times in each population (*i.e.*, we specified N_{learn} as $N_{learn} = 5$). Simulation results are summarized in Table 3. From Table 3, we can see that all the training patterns in Fig.11 were correctly classified by eight linguistic classification rules. The classification boundary by the eight linguistic classification rules is shown in Fig.12.

4.2 Simulation Results on Wine Data

Wine classification data [3] consist of 178 instances that were classified into three classes. Each instance has thirteen real-valued attributes. Corcoran & Sen [2] reported the following results of their genetic-based machine learning system with 60 non-fuzzy rules in each individual, 1500 individuals in each population, and 300 generations (*i.e.*, $1500 \times 300 = 450,000$ rule sets with 60 rules were examined in each trial):

> Best result: 100%,
> Average result: 99.5%,
> Worst result: 98.3%.

These results were classification rates for training data obtained by ten independent trials.

We applied our fuzzy classifier system to the same data. In our fuzzy classifier system, we used the following linguistic classification rule:

$$R_j : \text{If } x_1 \text{ is } A_{j1} \text{ and } \cdots \text{ and } x_{13} \text{ is } A_{j13} \text{ then Class } C_j \text{ with } CF_j .$$

The total number of linguistic classification rules is $6^{13} \approx 1.3 \times 10^{10}$. Our problem is to find a compact rule set with high classification performance from the 6^{13} linguistic classification rules.

In our fuzzy classifier system, we specified parameter values as follows:

> Population size: $N_{pop} = 60$,
> Weights in the fitness function: $w_{NCP} = 1$, $w_{NMP} = 5$,
> Crossover probability: 1.0,
> Mutation probability: 0.1,
> Number of replaced rules in each population: $N_{rep} = 12$,
> Iterations of the learning procedure in each generation: $N_{learn} = 5$,
> Learning rates in the learning procedure: $\eta_1 = 0.001$, $\eta_2 = 0.1$,
> Stopping condition: 500 generations.

From these parameter specifications, we can see that 500 rule sets with 60 linguistic classification rules were examined in our fuzzy classifier system. It should be noted that the total number of examined rule sets in our fuzzy classifier system (*i.e.*, 500 rule sets) was much smaller than that in the genetic-based machine learning system in Corcoran & Sen (*i.e.*, 450,000 rule sets).

We applied our fuzzy classifier system to the wine data 20 times, and a 100% classification rate was obtained in all the 20 trials. The average number of generations required for a 100% classification rate was 190.35. This means that our fuzzy classifier system found a rule set that can correctly classify all the training data after examining 190 rule sets on the average.

In Fig.13, we show the average performance of our fuzzy classifier system over the 20 trials. In Fig.13, $r(t)$ is the average classification rate at the t-th generation, and $r*(t)$ is the average best result obtained until the t-th generation: $r*(t) = \max\{r(1), r(2), ..., r(t)\}$. From this figure, we can see that almost all the training data were correctly classified after 200 generations while the classification rate at the initial generation was less than 10%.

From the comparison between our results and the above-mentioned results in [2], we can see that our approach found better rule sets by much smaller search effort than the genetic-based machine learning system in [2].

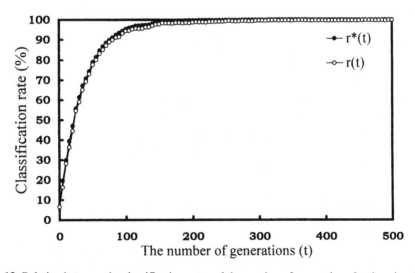

Fig.13 Relation between the classification rate and the number of generations for the wine data

4.3 Performance Evaluation for Test Data

In the previous subsection, we demonstrated the high performance of our fuzzy classifier system for training data. In this subsection, we examine the performance of our fuzzy classifier system for test data (*i.e.*, the generalization ability of our fuzzy classifier system). In computer simulations, the generalization ability was evaluated by the leaving-one-out procedure on appendicitis data and by the random resampling procedure (70% training patterns and 30% test patterns) on cancer data in the same manner as in Weiss and Kulikowski [22] where the generalization ability of various non-fuzzy classification methods was evaluated.

We applied our fuzzy classifier system with the following parameter values to these two data sets:

Population size: $N_{pop} = 100$,
Weights in the fitness function: $w_{NCP} = 1$, $w_{NMP} = 5$,
Crossover probability: 1.0,
Mutation probability: 0.1,
Number of replaced rules in each population: $N_{rep} = 20$,
Iterations of the learning procedure in each generation: $N_{learn} = 0$,
Stopping condition: 500 generations for appendicitis data,
　　　　　　　　　1000 generations for cancer data.

The appendicitis data include 106 patients and eight diagnostic tests [22]. But one test with some missing values was not used in our computer simulations as in Weiss and Kulikowski [22]. Thus we treated the appendicitis data as a seven-dimensional classification problem with two classes. The performance of our fuzzy classifier system for the appendicitis data was evaluated by the leaving-one-out procedure where the best population with the maximum classification rate for training data among 500 generations was used for classifying a test pattern in each trial in the leaving-one-out procedure. Simulation results are shown in Table 4. In Table 4, we also show the results of the linear discriminant, the nearest neighbor method and neural networks with the back-propagation (BP) algorithm reported in Weiss and Kulikowski [22] where various parameter specifications about the number of hidden units and the number of epochs were examined for the BP algorithm. From Table 4, we can see that the fuzzy classifier system outperformed the nearest neighbor method and slightly outperformed the linear discriminant and the BP algorithm.

Table 4 Performance of each classification method for the appendicitis data. The last three rows are the results in Weiss and Kulikowski [22].

	Error (E)	Correct (C)	Reject (R)	$C/(E+C)$
Fuzzy Classifier System	12.3%	86.8%	0.9%	87.6%
Linear Discriminant	13.2%	86.8%	0.0%	86.8%
Nearest Neighbor Method	17.9%	82.1%	0.0%	82.1%
Neural Networks with BP	14.2%	85.8%	0.0%	85.8%

The cancer data include 286 samples and nine tests [22]. Thus the cancer data are a nine-dimensional classification problems with two classes. Because the cancer data have some discrete attributes, we slightly modified our fuzzy classifier system. For example, we did not use three linguistic values (*i.e.*, MS: *medium small*, M: *medium*, and ML: *medium large*) for binary attributes with attribute values {0,1}. This is because the membership functions for these linguistic values in Fig.5 are 0 at the binary values {0,1}. From the same reason, we did not use MS (*medium small*) and ML (*medium large*) for ternary attributes with attribute values {0, 0.5, 1}. The performance of the fuzzy classifier system for the cancer data was evaluated by ten trials of the random resampling procedure where the best population with the maximum classification rate for training data among 1000 generations was used for classifying test patterns in each trial. The simulation results are summarized in Table 5. From Table 5, we can see that the fuzzy classifier system outperformed the nearest neighbor method and was comparable to the linear discriminant and the BP algorithm.

Table 5 Performance of each classification method for the cancer data. The last three rows are the results in Weiss and Kulikowski [22].

	Error (E)	Correct (C)	Reject (R)	$C/(E+C)$
Fuzzy Classifier System	22.0%	70.3%	7.7%	76.2%
Linear Discriminant	29.4%	70.6%	0.0%	70.6%
Nearest Neighbor Method	34.7%	65.3%	0.0%	65.3%
Neural Networks with BP	28.5%	71.5%	0.0%	71.5%

4.4 Discussions on Parameter Specifications in the Fuzzy Classifier System

Our fuzzy classifier system involves some parameters that should be pre-specified before we apply it to a particular pattern specification problem in hand. In this subsection, we examine the sensitivity of the performance of the fuzzy classifier system to the parameter specifications.

First we examined the relation between the performance of the fuzzy classifier system and the mutation probability. In the computer simulations in the previous subsections, we used a relatively large value as the mutation probability (*i.e.*, 0.1). We also performed computer simulations for the wine data with various mutation probabilities (*i.e.*, 0.001, 0.005, 0.01, 0.05, 0.1, 0.5, 1.0) in the same manner as in Subsection 4.2. It should be noted that the fuzzy classifier system with the mutation probability 1.0 can be viewed as a random search. Simulation results are summarized in Table 6. From Table 6, we can see that too small probabilities as well as too large probabilities led to poor performance of our fuzzy classifier system.

Table 6 Simulation results for the wine data with various mutation probabilities.

Mutation probability	0.001	0.005	0.01	0.05	0.1	0.5	1.0
Classification rate	64.0%	85.8%	97.2%	100%	100%	97.6%	40.2%

Next we examined the effect of the learning procedure on the performance of the fuzzy classifier system. In Subsection 4.2, we specified the number of iterations of the learning procedure in each generation as five (*i.e.*, $N_{learn} = 5$) and obtained a 100% classification rate for the training data in all the 20 trials. We also performed the same computer simulations for the wine data with no learning procedure (*i.e.*, $N_{learn} = 0$). From these computer simulations, we had the following results:

Best result: 100%,

Average result: 99.4%,

Worst result: 98.3%.

From these results, we can see that the learning procedure improved the classification rate for the training data.

In the learning procedure, the learning rates have a large effect on the performance. Throughout this paper, we specified the learning rates as $\eta_1 = 0.001$ and $\eta_2 = 0.1$. These values were suggested in Nozaki *et al.*[15] where various specifications of the learning rates were examined.

Finally we examined the effect of the weight values in the fitness function (*i.e.*, w_{NCP} and w_{NMP}) on the performance of the fuzzy classifier system. In Subsection 4.2, we performed the computer simulations for the wine data by specifying the weight values as $w_{NCP} = 1$ and $w_{NMP} = 5$. We also performed computer simulations for the wine data with various weight values. Simulation results are summarized in Table 7. From Table 7, we can see that high classification rates were obtained by a wide range of the weight values.

Table 7 Simulation results for the wine data with various weight values.

Weight values (w_{NCP}, w_{NMP})	(1, 0)	(1, 1)	(1, 2)	(1, 5)	(1, 10)
Classification rate	100%	100%	100%	100%	100%

Table 8 Simulation results for the cancer data with various weight values (training data).

Weight values (w_{NCP}, w_{NMP})	Error (E)	Correct (C)	Reject (R)	$C/(E+C)$
(1, 0)	15.8%	84.1%	0.1%	84.2%
(1, 1)	11.6%	86.8%	1.6%	88.2%
(1, 2)	12.4%	85.8%	1.9%	87.4%
(1, 5)	12.4%	81.6%	6.1%	86.8%
(1, 10)	12.2%	81.9%	6.0%	87.0%

Table 9 Simulation results for the cancer data with various weight values (test data).

Weight values (w_{NCP}, w_{NMP})	Error (E)	Correct (C)	Reject (R)	$C/(E+C)$
(1, 0)	26.2%	69.9%	4.0%	72.7%
(1, 1)	28.3%	66.5%	5.2%	70.1%
(1, 2)	26.6%	69.0%	4.4%	72.2%
(1, 5)	22.0%	70.3%	7.7%	76.2%
(1, 10)	23.5%	66.9%	9.7%	74.0%

As shown in Table 7, the weight value w_{NMP} (*i.e.*, penalty for misclassification) had no effect on the final results of the fuzzy classification system for the wine data. This is because the error rate was very low during the execution of the fuzzy classifier system for the wine data. We also performed computer simulations on the cancer data by using various weight values in the same manner as in Subsection 4.3. Simulation results are summarized in Table 8 for training data and Table 9 for test data. From these table, we can see that the weight value w_{NMP} for the misclassification has an effect on the performance of the fuzzy classifier system when the error rate is not negligible.

5. Conclusion

In this paper, we proposed a fuzzy classifier system for automatically generating linguistic classification rules for multi-dimensional pattern classification problems with many features. The performance of our fuzzy classifier system was examined by the application to the wine classification data with thirteen attributes. The high performance of our approach with linguistic classification rules was clearly demonstrated by computer simulations in comparison with the genetic-based machine learning system with non-fuzzy classification rules in Corcoran & Sen [2]. That is, our fuzzy classifier system found rule sets that can correctly classify all the training data after examining only 190 rule sets with 60 rules on the average. The performance of our fuzzy classifier system for test data was also examined by the leaving-one-out procedure and the random resampling procedure on real-world pattern classification problems. Simulation results indicated that our fuzzy classifier system outperformed the nearest neighbor method and was comparable to the linear discriminant and neural networks with the back-propagation algorithm. One advantage of the fuzzy classifier system over other classification methods is that classification knowledge is obtained as linguistic classification rules, each of which is easily understood by human users.

References

1. Bonarini, A.: Evolutionary learning of general fuzzy rules with biased evaluation functions: Competition and cooperation, *Proc. of 1st IEEE Conf. on Evolutionary Computation* (1994) 51-56.
2. Corcoran, A.L. and Sen, S.: Using real valued genetic algorithms to evolve rule sets for classification, *Proc. of 1st ICEC* (1994) 120-124.
3. Forina, M. *et al.*: Wine recognition data, Available via anonymous ftp from *ics.uci.edu* in directory */pub/machine-learning-databases/wine* (1992).
4. Goldberg, D.E.: *Genetic Algorithms in Search, Optimization, and Machine Learning*, Addison-Wesley, Reading, Massachusetts (1989).
5. Holland, J.H.: *Adaptation in Natural and Artificial Systems*, University of Michigan Press, Ann Arbor, Michigan (1975).
6. Ishibuchi, H., Nozaki, K. and Tanaka, H.: Distributed representation of fuzzy rules and its application to pattern classification, *Fuzzy Sets and Systems* **52** (1992) 21-32.

7. Ishibuchi, H., Nozaki, K., Yamamoto, N. and Tanaka, H.: Selecting fuzzy if-then rules for classification problems using genetic algorithms, *IEEE Trans. on Fuzzy Systems* **3** (1995) 260-270.

8. Ishibuchi, H., Nozaki, K., Yamamoto, N. and Tanaka, H.: Construction of fuzzy classification systems with rectangular fuzzy rules using genetic algorithms, *Fuzzy Sets and Systems* **65** (1994) 237-253.

9. Jang, J.S.R. and Sun, C.T.: Functional equivalence between radial basis function networks and fuzzy inference systems, *IEEE Trans. on Neural Networks* **4** (1993) 156-163 .

10. Karr, C.L.: Design of an adaptive fuzzy logic controller using a genetic algorithm, *Proc. of 4th ICGA* (1991) 450-457.

11. Karr, C.L. and Gentry, E.J.: Fuzzy control of pH using genetic algorithms, *IEEE Trans. on Fuzzy Systems* **1** (1993) 46-53.

12. Lee, C.C.: Fuzzy logic in control systems: fuzzy logic controller, *IEEE Trans. on Systems, Man and Cybernetics* **20** (1990) 404-435.

13. Nie, J. and Linkens, D.A.: Learning control using fuzzified self-organizing radial basis function network, *IEEE Trans. on Fuzzy Systems* **1** (1993) 280-287.

14. Nomura, H., Hayashi, I. and Wakami, N.: A self-tuning method of fuzzy reasoning by genetic algorithm, *Proc. of 1992 International Fuzzy Systems and Intelligent Control Conference* (1992) 236-245.

15. Nozaki, K., Ishibuchi, H. and Tanaka, H.: Adaptive Fuzzy-Rule-Based Classification Systems, IEEE Trans. on Fuzzy Systems (to appear).

16. Sugeno, M.: An introductory survey of fuzzy control, *Information Sciences* **36** (1985) 59-83.

17. Sugeno, M. and Yasukawa, T.: A fuzzy-logic-based approach to qualitative modeling, *IEEE Trans. on Fuzzy Systems* **1** (1993) 7-31.

18. Takagi, T. and Sugeno, M.: Fuzzy identification of systems and its applications to modeling and control, *IEEE Trans. on Systems, Man and Cybernetics* **15** (1985) 116-132.

19. Thrift, P.: Fuzzy logic synthesis with genetic algorithms, *Proc. of 4th ICGA* (1991) 509-513.

20. Valenzuela-Rendon, M.: The fuzzy classifier system: a classifier system for continuously varying variables, *Proc. of 4th ICGA* (1991) 346-353.

21. Wang, L.X. and Mendel, J.M.: Generating fuzzy rules by learning from examples, *IEEE Trans. on Systems, Man and Cybernetics* **22** (1992) 1414-1427.

22. Weiss, S.M. and Kulikowski, C.A.: *Computer Systems That Learn*, Morgan Kaufmann, San Mateo (1991).

Numerical Coding and Unfair Average Crossover in GA for Fuzzy Rule Extraction in Dynamic Environments

Tatsuya Nomura[1] and Tsutomu Miyoshi[2]

[1] ATR Human Information Processing Research Laboratories,
2-2, Hikaridai, Seika-cho, Soraku-gun, Kyoto 619-02, Japan
[2] Image Media Research Laboratories, SHARP CORPORATION,
1-9-2, Nakase, Mihama-ku, Chiba-shi, Chiba 261, Japan

Abstract. In this paper, we propose a GA with a new crossover method appropriate for real value chromosomes, called the "Unfair Average Crossover", an automatic fuzzy rule extraction method that uses our GA and a real value chromosome coding method in which parameters in membership functions of fuzzy if-then rules are directly represented. It is shown that our method is superior to conventional methods using discrete chromosome coding in cases where there is a tendency for data to change dynamically.

1 Introduction

Though fuzzy inference rules have usually been constructed through trial and error by humans, many methods with machine learning such as neural networks and genetic algorithms have recently been proposed for automatic rule extraction from a given set of input-output data examples. In particular, a number of fuzzy rule extraction methods with genetic algorithms (GA) have recently been proposed [1][2][7]. Most of them aim at selecting the most appropriate rules for partitioning the input data among the rules given in advance, or determining the numerical values of then-parts in simplified fuzzy inference rules. These methods use a discrete chromosome coding method such as binary coding, to represent the labels of the selected membership functions or the determined numerical values of the then-parts. From the viewpoint of fuzzy clustering of the input space, however, here are difficulties in adjusting the parameters of fuzzy membership functions strictly according to the distribution of the data. Moreover, they use one or multi-point crossover for bit strings and have high costs for the crossover and for the calculation of fitness values of individuals. Although some GAs use numerical coding and an average operator for crossover [3], this does not maintain the variety in the population.

In this paper, we propose a new crossover called the "Unfair Average Crossover" which is suitable for fuzzy clustering using numerical coding, and a new genetic algorithm. We also propose a method for automatic fuzzy rule extraction with our GA. Furthermore, we present results of simulations comparing with a conventional GA in cases where there is a tendency for data to change dynamically.

2 Numerical Coding and Unfair Average Crossover in GA for Fuzzy Rule Extraction

2.1 Representation of Fuzzy Rule

Our purpose in this paper is to cluster n-dimensional numerical vectors with fuzzy sets. We assume that each numerical vector is labeled with a discrete or symbolic value. When there are L labels in total, we prepare the following n-dimensional Gaussian fuzzy membership functions for each label:

$$F_{ki}(x_1,\ldots,x_n) = \prod_{j=1}^{n} F_{kij}(x_j) = \exp\left(-\frac{1}{2}\sum_{j=1}^{n} \frac{(\mu_{kij} - x_j)^2}{\sigma_{kij}{}^2}\right) \tag{1}$$

$$F_{kij}(x) = \exp\left(-\frac{1}{2}\sum_{j=1}^{n} \frac{(\mu_{kij} - x_j)^2}{\sigma_{kij}{}^2}\right) \tag{2}$$

$$(i = 1,\ldots,m,\ j = 1,\ldots,n,\ k = 1,\ldots,L)$$

Thus, we use m fuzzy sets to cluster the numerical vectors with the k-th label.

2.2 Numerical Coding

In this framework for fuzzy clustering, conventional GAs represent chromosomes with bit strings. These bit strings correspond to discrete labels of the membership functions given in advance, or discretized codes of numerical parameters from the membership functions in the if-parts of the rules. Conventional GAs execute bit operators for these bit strings and calculate the fitness values of all individuals based on the degree to which vectors are included in the fuzzy sets corresponding to the chromosomes. Even if these bit strings correspond to the numerical parameters, the conventional GAs do not deal with them as numerical values.

In this paper, we directly code the chromosome with an array of numerical parameters from the membership functions:

$$Ind_{ki} = (\mu_{ki1}\ \sigma_{ki1}\ \mu_{ki2}\ \sigma_{ki2}\ \cdots\ \mu_{kin}\ \sigma_{kin}) \quad (i = 1,\ldots,m,\ k = 1,\ldots,L) \tag{3}$$

Then, we construct one population POP_k with m individuals $\{Ind_{ki} : i = 1,\ldots,m\}$, a total of L populations POP_k ($k = 1,\ldots,L$), and independently execute genetic operators for each population. POP_k consists of all the chromosomes corresponding to the fuzzy membership functions for clustering the vectors with the k-th label. We also define the fitness value of the chromosome Ind_{ki} based on the degree to which the vectors are included in the corresponding fuzzy set F_{ki}, and select the individuals appropriate for fuzzy clustering of the input vectors.

2.3   Unfair Average Crossover

In the framework of conventional GAs, the above array is dealt with as bit strings made of discretized codes of each numerical parameter and the crossover operator is defined as a one-point or multi-point replacement between two parent bit strings. In some GAs, the crossover is an average operator, *i.e.*, one offspring array is made by setting each value of the offspring array to the average value of two elements at the corresponding location in the parent s'numerical arrays [3]. From the viewpoint of fuzzy clustering, these crossover have a problem in that the shapes of the appropriate fuzzy membership functions are not inherited by the subsequent generations.

We propose the Unfair Average Crossover (UFAC) to overcome the above problem. In the above average crossover, one offspring is made from two parents. In our UFAC, in conrast, two offsprings are made from two parents.

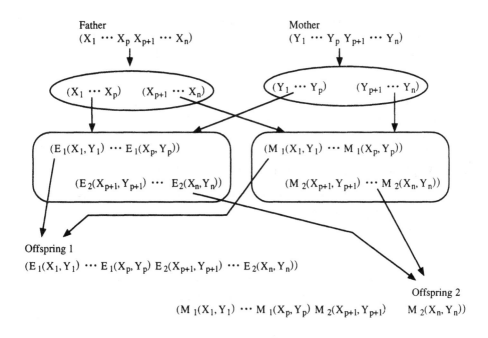

Fig. 1. Unfair Average Crossover

We make four kinds of numerical values $E_1(f, m)$, $E_2(f, m)$, $M_1(f, m)$, and $M_2(f, m)$ from two elements f, m at each location in the two parent arrays in the following way:

$$\left.\begin{array}{l}E_1(f,m) = (1+(1/a))f - (1/a)m \\ E_2(f,m) = -(1/a)f + (1+(1/a))m \\ M_1(f,m) = (1-(1/a))f + (1/a)m \\ M_2(f,m) = (1/a)f + (1-(1/a))m\end{array}\right\} \quad (a: \text{ real value}, a \geq 2) \quad (4)$$

Here, we call the parameter a in equation (4) the heritability. Note that the word "heritability" is an analogy from "narrow sence heritability" in quantitative genetics [12], although their strict definitions differ. $E_1(f,m)$ is the value far from the average in the direction of the father's value, $M_1(f,m)$ is the value near to the average from the direction of the father's value, $E_2(f,m)$ is far from the average in the direction of the mother's value, and $M_2(f,m)$ is near to the average from the direction of the mother's value.

Figure 1 shows the procedure of UFAC. We cut two parent arrays at one point and make two offspring arrays using E_1, E_2, M_1, and M_2. When we regard the parameters in equation (3) as a quantitative character of an individual, offspring 1 heavily inherits the characters of the parents and offspring 2 inherits an average amount.

Figure 2 shows the transition of shapes of membership functions by the average crossover and UFAC. For the average crossover, even if one of the parents has an appropriate character, the offspring blurs the character. In the case of UFAC, if one of the parents has an appropriate character, one of the offsprings inherits it to some degree, though the membership functions in the offsprings differ from those of the parents in the strict sense, dependent on the heritability. Furthermore, variety in the population is maintained by producing offsprings far from the average.

2.4 Application for Rule Extraction

We apply the above numerical chromosome coding and UFAC to extract fuzzy rules from a set of input-output data. We deal with the same data structure that Fuzzy ID3 [4][5] deals with:

$$\left.\begin{array}{l}S = \{S^{(1)}, S^{(2)}, \ldots, S^{(N)}\} \\ S^{(l)} = (a_{l1}, a_{l2}, \ldots, a_{ln} \; ; \; c_l), \\ a_l = (a_{l1}, a_{l2}, \ldots, a_{ln}) \in \mathbf{R}^n \\ c_l \in C = \{C_1, C_2, \ldots, C_L\} \quad (l = 1, \ldots, N) \\ S_k = \{S^{(l)} \in S : c_l = C_k\} \quad (k = 1, \ldots, L)\end{array}\right\} \quad (5)$$

We regard the above input-output data $S^{(l)}$ as the numerical vector a_l with label c_l and apply the fuzzy clustering with our GA. Finally, we get the following fuzzy if-then rules:

$$\text{If } I_1 \text{ is } F_{ki1} \text{ and } I_2 \text{ is } F_{ki2} \text{ and } \ldots \text{ and } I_n \text{ is } F_{kin},$$
$$\text{Then } c \text{ is } C_k \text{ with weight } \omega_{ki} \quad (6)$$

$$\sum_{i=1}^{m} \omega_{ki} = 1 \quad (k = 1, \ldots, L, \; i = 1, \ldots, m)$$

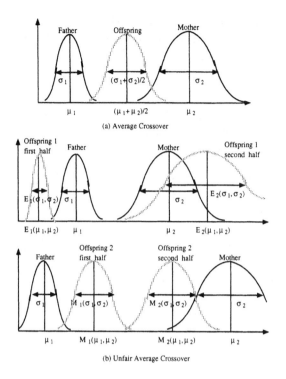

(a) Average Crossover

(b) Unfair Average Crossover

Fig. 2. Fuzzy Membership Functions of Parents and Offsprings for Average Crossover and Unfair Average Crossover

These rules with Gaussian membership functions do product-sum simplified inference with a degree of confidence and determine the degree where the output for a data with the numerical input attribute $I = (I_1, I_2, \ldots, I_n)$ is C_k for each k, $P_k(I)$, in the following way:

$$P_k(I) = \frac{p_k(I)}{\sum_{r=1}^{L} p_r(I)} \tag{7}$$

$$p_k(I) = \sum_{i=1}^{m} F_{ki}(I)\omega_{ki} \quad (k = 1, \ldots, L)$$

Here, $F_{ki}(I)$ is the value of the n-dimensional Gaussian fuzzy membership function for I in equation (1) and means the degree of fitness of I for the if-part of the rule in equation (6). $p_{ki}(I)$ means the weighted sum of the degree of fitness for the if-part and the degree of confidence of the rule. We get the result of inference $P_{ki}(I)$ by normalization of $p_{ki}(I)$.

Figure 3 shows the basic structure of automatic fuzzy rule extraction using our GA. We execute fuzzy clustering for the numerical vectors a_l in equation (5) with the chromosome Ind_{kl} in equation (refind), and then assign the final membership functions to the fuzzy parameters F_{ki} of the rules in equation (6) and the normalized fitness values to the degree of confidence of rules ω_{ki} in equation (6).

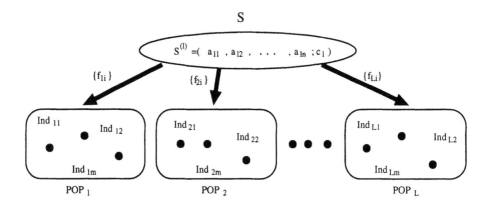

Fig. 3. Rule Extraction from Input-Output Data S by GA

We define the evaluation value e_{ki} and the fitness value f_{ki} of the chromosome Ind_{ki} in the following way:

$$e_{ki} = (1/2) \sum_{l=1}^{N} \left(F_{ki}(a_l) - T_k(c_l)\right)^2, \quad T_k(c_l) = \delta_{kl} \tag{8}$$

$$f_{ki} = \max_{r=1,\ldots,m}(e_{kr}) - e_{ki} + \alpha \min_{r=1,\ldots,m}(e_{kr}) \tag{9}$$

$$(k = 1,\ldots,L, \; i = 1,\ldots,m)$$

Here, $\alpha(> 0)$ is the selection pressure, e_{ki} in equation (8) means the squared error of the if-part of the corresponding rule for all the input-output data.

We execute rule extraction with the following procedures:

0 Initialize the arrays of real values in equation (3) randomly.
1 In each population POP_k, do roulette selection based on the fitness value in equation (9), unfair average crossover, and mutation by exchanging a randomly selected element in the array for a random value. (One generation)
2 Repeat 1 T times (T is a given integer)
3 Determine the degree of confidence in the rules by normalizing the final fitness values in the following way:

$$\omega_{ki} = \frac{f_{ki}}{\sum_{d=1}^{m} f_{kd}} \quad (k = 1,\ldots,L, \; i = 1,\ldots,m) \tag{10}$$

3 Simulation and Evaluation in Dynamical Environments

In this section, we show the results of simulations on automatic and adaptive fuzzy rule extraction to present the effectiveness of our GA in Section 2 by a comparison with conventional GAs. We executed the experiments by using data with three input attributes and three kinds of output values (in equation (5), $n = L = 3$).

3.1 Comparative methods

As comparative methods, we used the Simple GA with discrete chromosome coding and a GA with real value chromosome coding and an average crossover. In case of the Simple GA, we coded each real value parameter in the chromosome in equation (3) with a 64-bit binary or Gray code and the chromosome with a $64 \times 2 \times n$ length bit string. Then, we did roulette selection based on the fitness value in equation (9), a one-point crossover, and mutation with bit flip in each population POP_k individually. For the GA with real value chromosome coding and an average crossover, we used the method defined by exchanging UFAC in our GA defined in 2.4 for the average crossover.

3.2 Simulation 1

From the first to the 300th generation, we applied the GAs to those data with the distributions of input attribute vectors shown in Figure 4(a). Then, we assumed that the distributions drastically changed as shown in Figure 4(b) and applied the GAs to those data with these new distributions from the 301st to the 600th generation. Each distribution was a Gaussian distribution with width 1. For each distribution shown in Figure 4, we prepared 30 learning data and 30 testing data for one output value, for a total of 90 learning data and 90 testing data. We applied the GAs to the learning data, extracted the fuzzy rules, and evaluated the correctness rate of the rules for the learning data and the testing data at the 300th generation and the 600th generation.

Moreover, we formulated 12 kinds of sets of genetic parameters as shown in Table 1 and made an evaluation for each set of parameters. In each GA, the selection pressure α was 0.1 and the size of each population POP_k (m in equation (1) was 20 (*i.e.*, Total number of rules = 20×3).

Table 1. The sets of parameters used in Simulation 1

para No.	1	2	3	4	5	6	7	8	9	10	11	12
crossover rate	0.6	0.8	1.0	0.6	0.6	0.8	0.6	0.8	1.0	0.6	0.6	0.8
mutation rate	0.01	0.01	0.01	0.1	0.5	0.1	0.01	0.01	0.01	0.1	0.5	0.1
elitistic strategy	not	not	not	not	not	not	used	used	used	used	used	used

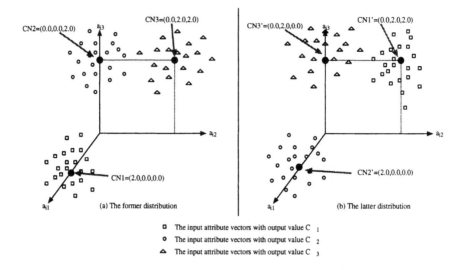

(a) The former distribution

(b) The latter distribution

□ The input attribute vectors with output value C_1
○ The input attribute vectors with output value C_2
△ The input attribute vectors with output value C_3

Fig. 4. Distributions of the Input Attribute Vectors in Simulation 1

The correctness rate of the rules is defined as:

$$C_r = 100 \times \frac{1}{3} \sum_{k=1,2,3} \frac{1}{30} \sum_{c_l=C_k} P_k(a_l) \tag{11}$$

Table 2 shows results of the simulations for the above data. Figures 5, 6, 7, 8, and 9 show the transition of the average of the evaluation values e_{ki} defined in equation (8) in each GA method for the above data. We show the tendency in the results of the correctness rates in the following:

Effect of crossover rate in cases where the elitistic strategy was not used: As shown in Table 2 (para No. 1) and Figure 5, when the crossover rate was low, the UFAC methods had 5-10% higher correctness rates for both the former and the latter distributions than the other methods and showed a high capacity of adaptation for environment changes. However, as shown in Table 2 (para No. 2, 3) and Figure 6, when the crossover rate was more than 0.8, the capacity of adaptation for environment changes was lower. The correctness rates of the Simple GA methods for the latter distribution were lower than those for the former distribution and the capacity of adaptation for environment changes was low independent of the crossover rate. The average crossover method had the same tendency as the Simple GA methods though its correctness rates were higher than those of the UFAC methods for a crossover rate 0.8.

Effect of crossover rate in cases where the elitistic strategy was used: As shown in Table 2 (para No. 8) and Figure 7, when the crossover rate was

Table 2. The Correctness Rates(%) of Each GA Method by the parameter sets in Table 1 for the Experimental Data in Simulation 1 (SGA-b: Simple GA using binary code, SGA-g: Simple GA using Graycode, RGA-av: GA using real value coding and average crossover, RGA-uav3: GA using real value coding and UFAC with a = 3.0, RGA-uav4: GA using real value coding and UFAC with a = 4.0)

para No.	1				2				3			
	former dist.		latter dist.		former dist.		latter dist.		former dist.		latter dist.	
	learning data	testing data	learning data	testing data	learning data	testing data	learning data	testing data	learning data	testing data	learning data	testing data
SGA-b	60.75	61.44	39.26	40.68	63.48	63.62	36.75	37.72	64.43	62.88	40.69	40.66
SGA-g	64.29	61.30	54.96	54.61	57.07	56.89	32.33	32.68	54.95	54.49	43.43	43.98
RGA-av	64.67	65.41	52.83	52.47	72.53	73.92	53.00	53.85	68.94	69.75	54.91	56.03
RGA-uav3	69.65	70.90	61.35	61.55	66.85	67.66	53.85	55.12	69.35	70.61	62.72	62.50
RGA-uav4	69.97	71.35	69.96	70.62	70.06	71.43	57.18	57.46	66.36	67.49	52.45	51.71

para No.	4				5				6			
	former dist.		latter dist.		former dist.		latter dist.		former dist.		latter dist.	
	learning data	testing data	learning data	testing data	learning data	testing data	learning data	testing data	learning data	testing data	learning data	testing data
SGA-b	68.11	67.84	58.69	61.37	57.78	58.14	71.63	71.37	63.28	63.11	55.23	53.72
SGA-g	69.75	70.03	51.39	55.50	67.37	66.56	73.88	75.75	64.99	65.65	61.77	62.06
RGA-av	64.84	66.00	65.85	66.48	64.42	65.00	62.08	62.34	68.52	69.83	67.86	68.86
RGA-uav3	72.69	74.31	72.19	73.22	61.36	61.77	58.56	58.83	72.64	74.32	69.28	69.71
RGA-uav4	77.51	79.72	70.91	71.28	60.28	60.71	49.57	49.94	66.68	68.15	76.78	76.60

para No.	7				8				9			
	former dist.		latter dist.		former dist.		latter dist.		former dist.		latter dist.	
	learning data	testing data	learning data	testing data	learning data	testing data	learning data	testing data	learning data	testing data	learning data	testing data
SGA-b	54.81	53.67	33.19	33.50	59.51	59.48	32.09	32.66	52.71	52.81	35.08	36.25
SGA-g	59.34	56.90	49.51	50.57	63.13	62.55	18.03	16.75	59.83	60.44	31.81	32.01
RGA-av	62.50	63.98	56.41	55.83	72.82	74.05	34.06	34.08	67.15	69.68	63.77	63.47
RGA-uav3	71.54	72.97	60.07	61.54	76.27	77.67	74.90	75.40	73.62	74.79	58.59	60.48
RGA-uav4	70.59	71.92	67.14	68.04	69.36	70.74	72.47	73.26	68.74	69.80	66.89	67.13

para No.	10				11				12			
	former dist.		latter dist.		former dist.		latter dist.		former dist.		latter dist.	
	learning data	testing data	learning data	testing data	learning data	testing data	learning data	testing data	learning data	testing data	learning data	testing data
SGA-b	55.56	54.67	60.76	62.73	68.64	68.94	66.37	66.28	64.97	64.14	49.17	50.63
SGA-g	56.66	56.16	62.38	61.01	63.85	64.28	68.66	69.20	62.27	61.98	57.29	59.12
RGA-av	66.69	67.83	68.56	68.70	57.38	59.51	57.75	58.12	64.71	65.65	68.20	68.76
RGA-uav3	70.46	72.10	73.99	74.86	53.29	54.14	58.08	57.41	74.32	76.01	84.10	84.85
RGA-uav4	67.44	69.20	73.17	74.11	56.20	56.42	58.08	58.15	64.25	64.87	73.60	74.37

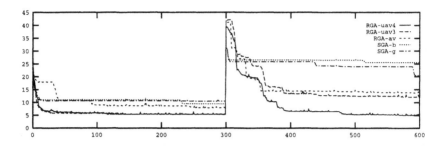

Fig. 5. Transition of the evaluation values of each GA by parameter set 1 in Table 1 for Simulation 1 (horizontal axis: generation number, vertical axis: the average of the evaluation values defined in equation (8))

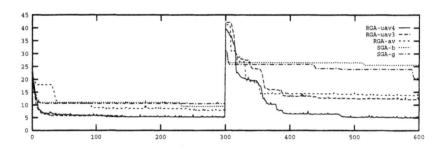

Fig. 6. Transition of the evaluation values of each GA by parameter set 2 in Table 1 for Simulation 1

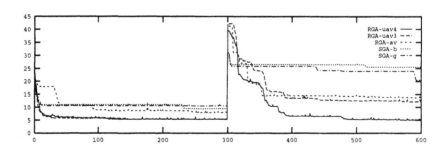

Fig. 7. Transition of the evaluation values of each GA by parameter set 8 in Table 1 for Simulation 1

0.8, the UFAC methods had higher correctness rates for both the former and the latter distributions than the other methods and showed a high capacity of adaptation for environment changes. Though the correctness rates of the average crossover method for the former distribution were as high as those of the UFAC methods, the correctness rates for the latter distribution were about 30% lower and the capacity of adaptation for environment changes was low. However, when the crossover rate was very high, the correctness rates of the average crossover method for the latter distribution became as high as those of the UFAC methods. Introducing the elitistic strategy in the Simple GA methods made the capacity of adaptation lower.

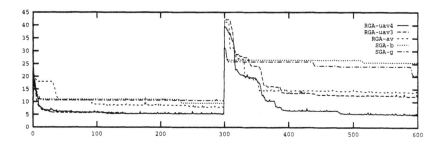

Fig. 8. Transition of the evaluation values of each GA by parameter set 4 in Table 1 for Simulation 1

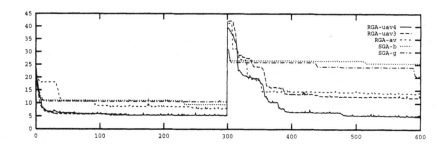

Fig. 9. Transition of the evaluation values of each GA by parameter set 5 in Table 1 for Simulation 1

Effect of mutation rate: As shown in Table 2 (para No. 4, 6) and Figure 8, when the mutation rate was 0.1, all of the methods showed a higher capacity of adaptation for environment changes. In particular, the correctness rates of the Simple GA method with binary code, the average crossover method, and the UFAC method with a = 3.0 for the latter distribution were about 10% higher. However, as shown in Table 2 (para No. 5) and Figure 9, when the mutation rate was very high, the correctness rates of the methods with real value coding were lower. On the other hand, the correctness rates of the Simple GA methods were about 10% higher than those of the others. As shown in Table 2 (para No. 10, 11, 12), a common tendency existed between both the case where the elitistic strategy was used and the case where it was not.

Effect of heritability for the UFAC method: As shown in Table 2 (para No. 1, 6, 7, 9), when the crossover rate was low, the UFAC method with heritability 3.0 had a lower correctness rate for the latter distributions than the same method with heritability 4.0. Identical results were obtained, both when the crossover and mutation rate were high and the elitistic strategy was not used, and when the crossover rate was too high, the mutation rate was low, and the elitistic strategy was used. However, as shown in Table 2 (para No. 3, 8, 12), the UFAC method with heritability 3.0 had a higher correctness rate for the former and latter distributions than the same method with heritability 4.0, both when the crossover rate was too high , the mutation rate was low and the elitistic strategy was not used, and when the crossover rate was high and the elitistic strategy was used.

3.3   Simulation 2

We assumed that the distributions of input attribute vectors gradually changed from the former to the latter distribution in Simulation 1 every 30 generations as shown in Table 3 and applied the GAs to those data with these distributions of the Gaussian type. For each distribution shown in Table 3, we prepared a total of 90 learning data and 90 testing data in the same way as in Simulation 1. We extracted the fuzzy rules from the learning data and evaluated the correctness rate of the rules for the testing data defined in equation (11) every 30 generations when the distribution of the input attribute vectors changed.

We made a comparison between the GAs for a crossover rate of 0.6 and a mutation rate of 0.01. Moreover, we made a comparison between the Simple GA methods with a crossover rate of 0.6 and a mutation rate of 0.5, the average crossover method and the UFAC methods with a crossover rate of 0.8 and a mutation rate of 0.1. The above parameter set in each method showed the highest correctness rates among the 12 kinds of parameter sets in Simulation 1.

Figure 10 (Upper) shows the transition of the correctness rates of each method for a crossover rate of 0.6 and a mutation rate of 0.01. Figure 10 (Lower) shows the transition of the correctness rates of the Simple GA method with binary code and Gray code for a crossover rate of 0.6 and a mutation rate of 0.5,

Table 3. Transition of the distributions of input attribute vectors in Simulation 2

gene-ration number	Center of Input attribute vectors with C_1	Center of Input attribute vectors with C_2	Center of Input attribute vectors with C_3	deviation of each distribution
1-30	(2, 0, 0)	(0, 0, 2)	(0, 2, 2)	1
31-60	(1.9, 0.1, 0.1)	(0.1, 0, 1.9)	(0, 1.9, 2)	0.95
61-90	(1.8, 0.2, 0.2)	(0.2, 0, 1.8)	(0, 1.8, 2)	0.9
91-120	(1.7, 0.3, 0.3)	(0.3, 0, 1.7)	(0, 1.7, 2)	0.85
121-150	(1.6, 0.4, 0.4)	(0.4, 0, 1.6)	(0, 1.6, 2)	0.8
151-180	(1.5, 0.5, 0.5)	(0.5, 0, 1.5)	(0, 1.5, 2)	0.75
181-210	(1.4, 0.6, 0.6)	(0.6, 0, 1.4)	(0, 1.4, 2)	0.7
211-240	(1.3, 0.7, 0.7)	(0.7, 0, 1.3)	(0, 1.3, 2)	0.65
241-270	(1.2, 0.8, 0.8)	(0.8, 0, 1.2)	(0, 1.2, 2)	0.6
271-300	(1.1, 0.9, 0.9)	(0.9, 0, 1.1)	(0, 1.1, 2)	0.55
301-330	(1, 1, 1)	(1, 0, 1)	(0, 1, 2)	0.5
331-360	(0.9, 1.1, 1.1)	(1.1, 0, 0.9)	(0, 0.9, 2)	0.52
361-390	(0.8, 1.2, 1.2)	(1.2, 0, 0.8)	(0, 0.8, 2)	0.54
391-420	(0.7, 1.3, 1.3)	(1.3, 0, 0.7)	(0, 0.7, 2)	0.56
421-450	(0.6, 1.4, 1.4)	(1.4, 0, 0.6)	(0, 0.6, 2)	0.58
451-480	(0.5, 1.5, 1.5)	(1.5, 0, 0.5)	(0, 0.5, 2)	0.61
481-510	(0.4, 1.6, 1.6)	(1.6, 0, 0.4)	(0, 0.4, 2)	0.68
511-540	(0.3, 1.7, 1.7)	(1.7, 0, 0.3)	(0, 0.3, 2)	0.76
541-570	(0.2, 1.8, 1.8)	(1.8, 0, 0.2)	(0, 0.2, 2)	0.84
571-600	(0.1, 1.9, 1.9)	(1.9, 0, 0.1)	(0, 0.1, 2)	0.92
601-630	(0, 2, 2)	(2, 0, 0)	(0, 0, 2)	1

the average crossover method and the UFAC method with a = 3.0 and a = 4.0, a crossover rate of 0.8 and a mutation rate of 0.1.

As shown in Figure 10 (Upper), when the mutation rate is low, the correctness rates of the UFAC methods change between 50% and 70%, though those of the average crossover method change between 40% and 60%, and those of the Simple GA methods eventually decrease to about 30%. In this case, the UFAC methods showed a higher capacity of adaptation for environment changes than the others. Moreover, the correctness rates of the UFAC method with heritability 3.0 in the latter steps were lower than those of the same method with heritability 4.0. As shown in Figure 10 (Lower), when the mutation rate is high, the correctness rates of the Simple GA method with Gray code change between 60% and 80%, and those of the UFAC methods change between 50% and 80%. However, those of the Simple GA method with binary code and the average crossover method change between 40% and 65%. In this case, the UFAC methods and the Simple GA method with Gray code showed a higher capacity of adaptation for environment changes than the others.

Figures 11, 12, 13, and 14 show the transition of the membership function F_{121} in each method for a crossover rate of 0.6 and a mutation rate of 0.01. As the center of the distribution of the input attribute vectors with label C_1 moved from (2.0, 0.0, 0.0) to (0.0, 2.0, 2.0), the membership functions in the UFAC methods moved in the same direction. Although that in the average crossover method also moved in the same direction, the width of the membership function did not

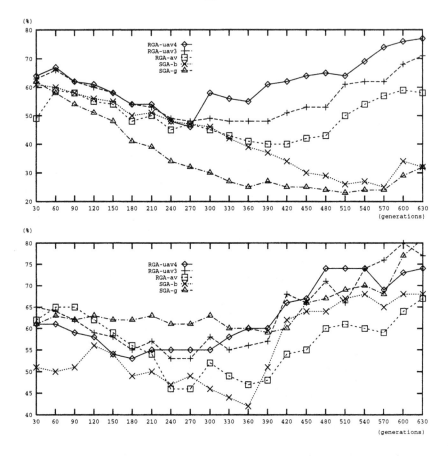

Fig. 10. Transition of the correctness rates of each GA method in Simulation 2 (Upper: Case of Crossover Rate = 0.6 and Mutation Rate = 0.01, Lower: Case of Crossover Rate = 0.6 and Mutation Rate = 0.5 for Simple GA, Crossover Rate = 0.8 and Mutation Rate = 0.1 for GA with Numerical Chromosomes)

match the distribution. The membership functions in the Simple GA methods hardly changed except for the width. Even the membership functions in the UFAC methods did not exactly reflect the distribution of the input attribute vectors in the strict sence, but their corrctness rates were higher than those of the other methods. As a reason for this result, we consider the usage of the evaluation value e_{ki} in equation (8), which is a simple squared error for data and does not exactly reflect the distribution of data.

3.4 Discussion

The following discussion is given based on the above results. When the crossover rate is too high or the heritability is low, UFAC produces many offsprings near

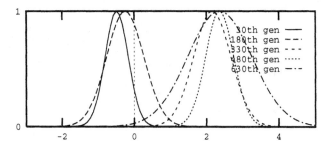

Fig. 11. The transition of the membership function F_{121} in the UFAC method with $a = 4.0$

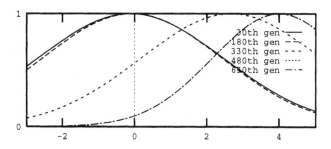

Fig. 12. The transition of the membership function F_{121} in the UFAC method with $a = 3.0$

the average of the chromosomes suitable for the recent environment and they survive under selection. Thus, the variety in the population becomes lower in the same way as the case using average crossover, and the rule extraction with the UFAC is not able to follow drastic changes in the environment. On the other hand, once the environment drastically changes and the fitness values of offsprings far from the average become higher, the elitistic strategy makes them survive. Moreover, in an environment with frequent changes, a high crossover rate produces offsprings appropriate for the changing environment. Thus, the rule extraction with UFAC has a higher capacity of adaptation for environment changes than the other methods. Because the average crossover method does not produce offsprings far from the average, the rule extraction does not have as high a capacity of adaptation as that with the UFAC, independent of elitistic strategy usage.

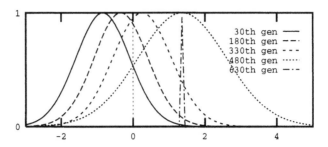

Fig. 13. The transition of the membership function F_{121} in the average crossover method

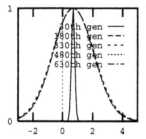 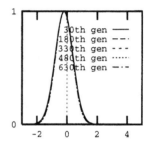

Fig. 14. The transition of the membership function F_{121} in the Simple GA method (Left: Binary coding, Right: Gray coding)

Moreover, mutation maintains the variety in the population under both real value coding and bit coding. In particular, in the case of the UFAC method, the combination of a high mutation rate, high heritability, and the use of an elitsitic strategy derives a higher capacity of adaptation for environmental changes. In the case of real value coding, however, a too high mutation rate causes an oscillation of the average of the evaluation values as shown in Figure 9 and the population does not converge. That is why we exchange a random value for the values in chromosomes for our mutation. On the other hand, in the case of bit coding, a high mutation rate is needed for adaptation for environment changes because a small population size often causes genetic drift.

Furthermore, Simple GA methods need much more time for the genetic process than GAs with real value chromosomes. Table 4 shows the time taken for a total of 600 generations in each method in the above simulations with a Sparc-

Station 5. As shown in Table 4, the time for the Simple GA methods was 300 times more than that of the GAs with real value chromosomes. That is why the calculation of the fitness values took more time; the time was needed for translation between the bit string of the chromosome and the numerical value used in calculating of the fitness.

Table 4. Time taken for a total of 600 generations in each method

Method	time
real value coding and UFAC	about 2 minutes
real value coding and average crossover	about 2 minutes
Simple GA with binary coding	about 10 hours
Simple GA with Gray coding	about 11 hours

4 Problems

Although we have proposed and verified numerical coding and unfair average crossover in GA for fuzzy rule extraction in the above sections, some theoretical problems exist in our method. We have not yet solved these problems and only describing them here. The first problem is behavior analysis of GA using numerical coding. For the Simple GAs, many results on theoretical behavior analysis to find convergence properties of populations after many generation alternations and conditions on crossover and mutation rates for the convergence have been reported. We have to present a similar mathematical analysis for GAs using numerical coding.

For GAs using bit strings, the set of all possible population states is finite and the theories on finite Markov chains can be used [8][9][10]. For GAs using numerical coding, the set of all possible population states is an infinite and uncountable set. As one approach, we can consider using the theories for stochastic processes on uncountable state spaces. Moreover, when each value in the numerical chromosomes in equation (3) are represented as a floatng point in a computer, the set of all possible population states is finite and we can apply the theories on finite Markov chains to our GA.

As another approach, we can consider using quantitative genetics [11][12] from the viewpoint of biology. By regarding the parameters in equation (3) as a quantitative character of an individual, the average and deviation of the character in a population can be analyzed using quantitative genetics.

The second problem is related to the fuzzy rules we use. The fuzzy if-then rules in equations (6) and (7) are not within the conventional definitions of indirect fuzzy inference. Thus, we have to present the theoretical framework for our rule formula and conditions in which the rules are effective.

The above two problems are both important and difficult. Although we cannot deal with them in this paper, they will be dealt with in future work.

5 Conclusion

We proposed the Unfair Average Crossover in a GA with real value chromosomes, and verified the effectiveness of our GA for fuzzy rule extraction by comparative simulations with conventional GAs. We concluded that our method is more effective than conventional methods such as Simple GA with bit string for automatic and adaptive fuzzy rule extraction in cases where the distribution of input attributes dynamically changes. Moreover, we could conclude that our GA requires much lower process time than the Simple GA with bit string.

Furthermore, we presented the current problems for our methods and discussed some approaches to solve them.

References

1. Nakanishi, S. : Fuzzy Control by Genetic Algorithms. SYSTEMS, CONTROL AND INFORMATION **38, 11** (1994) 613–618
2. Lee, M. A. : On Genetic Representation of High Dimensional Fuzzy Systems. Proc. ISUMA-NAFIPS'95 (1995) 752–757
3. Davis, L. : HANDBOOK OF GENETIC ALGORITHMS (1990) Van Nostrand Reinhold
4. Umano, H., Okamoto, H., Hatono, I., Tamura, H. : Generation of Fuzzy Rules from Numerical Data by ID3 Algorithm and Their Inference Method. Proc. 9th Fuzzy System Symposium (Sapporo) (1993) 858–860
5. Sakurai, S., Araki, D. : Generating a Fuzzy Decision Tree by Inductive Learning. T. IEE Japan **113-C, 7** (1993) 488–494
6. Wada, K : Foundations of Genetic Algorithm. Computer Today **47** (1992) 49–61
7. Valenzuela-Rendon, M. : The Fuzzy Classifier System: A Classifier System for Continuously Varying Variables, Proc. 4th ICGA (1991) 346–353
8. Davis, T., Principe, J. C. : A Markov Chain Framework for the Simple Genetic Algorithm. Evolutionary Computation **1, 3** (1993) 269–288
9. Vose, M. D., Liepins, G. E. : Punctuated Equilibria in Genetic Search. Complex Systems **5** (1991) 31–44
10. Dawid, H. : A Markov Chain Analysis of Genetic Algorithms with a State Dependent Fitness Function. Complex Systems **8** (1994) 407–417
11. Falconer, D. S. : Introduction to Quantitative Genetics (Third Edition). (1989) Longman Group UK Ltd.
12. Crow, J. F. : BASIC CONCEPTS IN POPULATION, QUANTITATIVE, AND EVOLUTIONARY GENETICS. (1986) W. H. FREEMAN AND COMPANY, New York

Acquisition of Fuzzy Rules from DNA Coding Method

Tomohiro Yoshikawa, Takeshi Furuhashi and Yoshiki Uchikawa

Department of Information Electronics, Nagoya University
Furo-cho, Chikusa-ku, Nagoya, 464-01, Japan
Tel.+81-52-789-2793 Fax. +81-52-789-3166
E-mail: yoshi@bioele.nuee.nagoya-u.ac.jp

ABSTRACT

This paper presents a new coding method based on biological DNA. A mechanism of development from the artificial DNA is also presented in this paper. This mechanism realizes flexible representation of fuzzy rules. The artificial DNA is composed of four kinds of bases. The proposed DNA allows redundancy and overlaps of genes. Fuzzy rules for mobile robots are acquired through chasing and avoiding operations. An application of virus and enzyme operators into the artificial DNA is also presented in this paper.

I. INTRODUCTION

Genetic Algorithms (GAs)[1][2] have been widely studied and applied to many problems. This paper proposes a new coding method based on biological DNA and a mechanism of development from the artificial DNA. This paper calls this coding method the "DNA coding method". The GA based on a biologically motivated model was proposed by Willfried Wienholt[3]. This model took into account gene expression which involves translation of nucleotide sequences of DNA into amino acid sequences. This method, however, is devised to solve parameter optimization problems.

Fuzzy controls, described in linguistic IF-THEN rules, have been widely used in industry for their high degree of performance in human-computer interactions. Demand for fuzzy inference systems which can describe complex, multi-input/output systems is growing. C.L.Karr[4][5] has proposed the application of the GA to the design of fuzzy logic controllers, and his work was a pioneering effort in the application of the GA to fuzzy controls. M.Valenzuela-Rendon[6] has proposed a fuzzy classifier system (FCS) by introducing fuzzy logic into the classifier system[7] and applying the FCS to approximate a nonlinear function. T.Furuhashi et al.[8-10] have studied the application of the FCS to knowledge finding for fuzzy controls. M.A.Lee and H.Takagi[11] have devised another interesting approach to the fusion of fuzzy logic and the GA. They have presented a method to control the parameters of the GA, i.e. mutation rate, crossover rate, etc., by fuzzy logic. Selection of input variables for fuzzy control rules was addressed by T.Hashiyama et al.[12] using an efficient bacterial GA[13].

This paper studies acquisition of effective fuzzy rules using the proposed DNA coding method. These fuzzy rules are used to control mobile robots which play chasing and avoiding. Selection of input variables and tuning of membership functions are done. This new coding method is compared with the coding method in [12], and the results show a higher performance of the proposed method. This paper also presents an application of virus and enzyme operators into the proposed DNA coding method.

II. DNA CODING METHOD

Fig.1(a) shows a flow from biological DNA to cells. The biological DNA consists of nucleotides which have four bases, Adenine(A), Guanine(G), Cytosine(C), Thymine(T)[14][15]. Most of these bases in the top figure in Fig.1(a) are not used for the synthesis of proteins. A messenger RNA (mRNA), which has many unused parts, is first synthesized from the DNA. In the synthesis of RNA, each base is translated into the complementary base i.e. T into A, G into C and so on. Moreover in the RNA the base U is used instead of T. Then the unused parts are cut out. This operation is a splicing. After this splicing has occurred, the mRNA is completed. Three successive bases called codons are allocated sequentially in the mRNA. These codons are the codes for amino acids. 64 kinds of codons correspond to 20 kinds of amino acids. The details of translation into amino acid from codons are omitted here. This allocation of amino acid makes proteins, and proteins make up cells.

Fig.1(b) shows the proposed DNA coding method and the flow of development to sets of fuzzy rules. This figure shows a correspondence of the proposed method to the biological development. The GA usually used a coding method specifically devised for each problem and it had no redundant parts. This conventional coding method could be regarded as a coding into complete mRNA. This paper presents a new coding method using the DNA itself and a way of development from the DNA to sets of fuzzy rules. A chromosome consists of combinations of four bases, A, G, C, T. The chromosome has many redundant parts, and after a splicing, the mRNA is completed. In this artificial RNA synthesis, each base is translated into the same base. The codons in Fig.1(b) also correspond to amino acids. Unlike the biological amino acid, each artificial amino acid has several meanings, and the meanings of a gene is determined by the combination of the amino acids. An amino acid can be translated as an input variable or a form of membership function, and so on. A sequence of amino acids makes a fuzzy rule. The DNA chromosome makes up sets of fuzzy rules for controlling a mobile robot.

Fig.2 shows an example of the DNA chromosome and its translation mechanism. In this figure a gene starts from the start codon ATG, and ends at the end codon TAG, and codons in the gene are translated into amino acids: Tyr, Thr, $\cdots$. Each amino acid has its own role for the problem.

By the proposed mechanism of development from the DNA, the starting point can be shifted from a base to another and some genes overlapping on other genes can be translated. Each overlapping gene plays an important role. Fig.3 shows this overlapped representation. In this figure, GENE5 in addition to GENE3 and GENE4 can be read

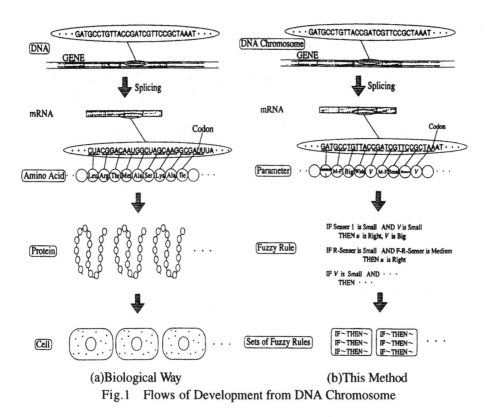

(a)Biological Way   (b)This Method

Fig.1 Flows of Development from DNA Chromosome

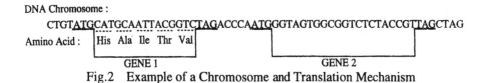

DNA Chromosome :

CTGTATGCATGCAATTACGGTCTAGACCCAATGGGTAGTGGCGGTCTCTACCGTTAGCTAG

Amino Acid :   His Ala Ile Thr Val

GENE 1   GENE 2

Fig.2 Example of a Chromosome and Translation Mechanism

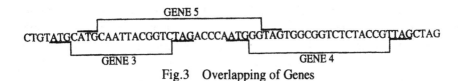

GENE 5

CTGTATGCATGCAATTACGGTCTAGACCCAATGGGTAGTGGCGGTCTCTACCGTTAGCTAG

GENE 3   GENE 4

Fig.3 Overlapping of Genes

from the DNA chromosome. This chromosome has redundancy and also compresses information by overlaps of genes.

Fig.4 shows examples of crossover and mutation. Fig4(a) is an example of one point crossover. Right hand sides from the crossover points are exchanged and new GENE2', 4', 5' are generated. By this method, depending on the crossover points, genes can be drastically changed. There is no constraint on the crossover points. Fig4(b) shows an example of mutation. One base indicated in the figure is changed from T to G. As a result, the GENE1 is changed to GENE1'. By this change, the start codon ATG is newly generated and new GENE7 is generated.

Fig.5 shows examples of virus and enzyme operations. Though the biological function of virus and enzyme is not known completely, this paper uses the virus and enzyme in the following way. Fig.5(a) shows a virus operation. A part of sequence of DNA from other DNA chromosome moves into the chromosome. As a result, the two genes of GENE7", GENE7"' are generated from GENE7. Fig.5(b) shows an enzyme operation. The enzyme distinguishes two amino acids, and splices the part between the two. In this figure the part between the codon AGT (Ser) and the codon GGT (Gly) is cut out, and GENE4' is deleted and GENE7"" is produced from GENE7.

The DNA coding method has the following features:

(a)Flexible representation of knowledge.

(b)The coding is redundant and overlapped.

(c)The length of the chromosome is variable.

(d)No constraint on crossover points.

III. PROBLEM FORMULATION FOR KNOWLEDGE ACQUISITION

The effects of the proposed method are demonstrated through the evolution of fuzzy rules developed from the artificial DNA. Fig.6 shows the simulation conditions of the problem and the construction of two robots. Two different types of mobile robots play chasing and avoiding in the area of 2.33m wide and 3m long surrounded by walls. There are several foods in the area for the avoiding robot. The radius of the chasing robot is 150mm, and that of the avoiding robot is 100mm. The chasing robot has eight ultra-sonic sensors (seven in the front, and one in the rear). These sensors can measure the distances between obstacles and themselves in the range of 200mm to 1700mm. The chasing robot must acquire rules to catch the other robot. The avoiding robot has twelve infrared sensors to see around. These sensors can measure limited distance less than 350mm. Therefore the avoiding robot cannot recognize the approaching robot until the enemy comes near to it. The avoiding robot must find fuzzy rules not to be caught by the other robot. The serial number of the front sensor of each robot is No.0. That of the next sensor in the counterclockwise side is No.1. Each robot has a chromosome containing a set of fuzzy rules. The fuzzy rules steer and accelerate/deccelerete the robot to chase/avoid the other robot and stay away from the walls.

The robot which reaches or avoids the other robot receives more payoffs from the

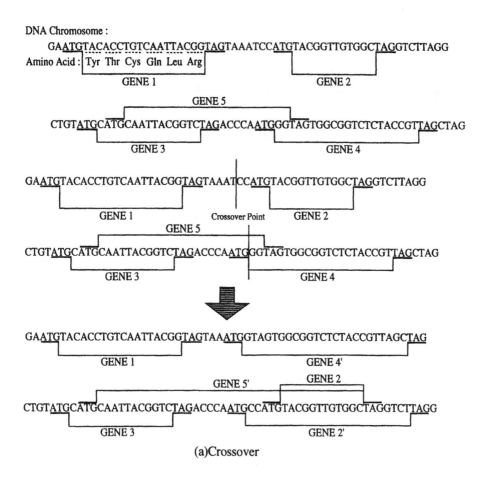

(a)Crossover

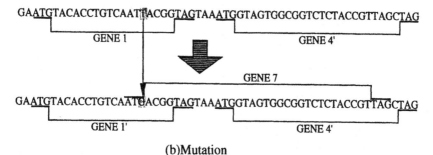

(b)Mutation

Fig.4   Crossover and Mutation

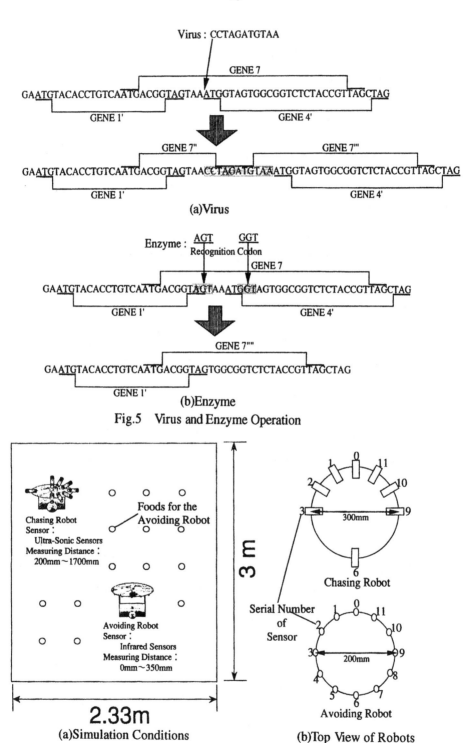

Virus : CCTAGATGTAA

GENE 7

GAATGTACACCTGTCAATGACGGTAGTAAATGGTAGTGGCGGTCTCTACCGTTAGCTAG

GENE 1'   GENE 4'

GENE 7"   GENE 7'''

GAATGTACACCTGTCAATGACGGTAGTAACCTAOATGTAAATGGTAGTGGCGGTCTCTACCGTTAGCTAG

GENE 1'   GENE 4'

(a)Virus

Enzyme : AGT   GGT
         Recognition Codon
                   GENE 7

GAATGTACACCTGTCAATGACGGTAGTAAATGGTAGTGGCGGTCTCTACCGTTAGCTAG

GENE 1'   GENE 4'

GENE 7""

GAATGTACACCTGTCAATGACGGTAGTGGCGGTCTCTACCGTTAGCTAG

GENE 1'

(b)Enzyme

Fig.5   Virus and Enzyme Operation

Foods for the
Avoiding Robot

Chasing Robot
Sensor :
 Ultra-Sonic Sensors
Measuring Distance :
 200mm~1700mm

3 m

Avoiding Robot
Sensor :
 Infrared Sensors
Measuring Distance :
 0mm~350mm

2.33m

(a)Simulation Conditions

300mm

Chasing Robot

Serial Number
of
Sensor

200mm

Avoiding Robot

(b)Top View of Robots

Fig.6   Simulation Conditions and Construction of Two Robots

environment. Considering these payoffs as fitness values, the genetic operators are applied to the chromosomes and the fuzzy rules are evolved.

IV. APPLICATION OF DNA CODING METHOD

This chapter applies the proposed DNA coding method to the fuzzy control rules of two mobile robots. The way of development from DNA chromosome to the fuzzy rules and the way to find rules are described in this section.

4.1 Representation of rules

The candidates for the input parameters of each robot are the detected values of sensors D and the robot's velocity V, and those for the output parameters are the steering angle u, and its velocity V. The chromosome has sets of fuzzy rules which are represented by IF~ THEN~ rules. The chromosome determines combinations of input/output parameters and membership functions of each fuzzy rules. The central position x_c and the width σ of the membership functions are also encoded into the chromosome.

In the biological DNA, a gene starts from the start codon ATG, and ends at the end codon TAA, TAG or TGA. In this paper, a gene also starts from the start codon ATG which corresponds to IF. The end codon is not definitely determined. A gene consists of the codons between IF codon and some related codons succeeding to THEN codon. When there is no rule in a chromosome, the robot having that chromosome moves only straight forward. Even if the lengths of all the chromosomes in population are too short to have a rule, the robots in population will get rules by the genetic operations i.e. crossover, mutation, and so on. Reading from the top of the DNA chromosome, translation to a fuzzy rule starts upon finding the start codon ATG. As described in Section II, overlaps of genes is allowed in the DNA chromosome. After reading a fuzzy rule, re-reading is restarted from the second base of the IF-codon and a new IF codon is sought. Fig.7 shows the flow of translation from the DNA chromosome into a fuzzy rule, and Table 1 shows the correspondence between the amino acids and the parameters defined in this paper. Like the biological process, the 64 kinds of codons correspond to 20 kinds of amino acids. The meanings of each amino acid is determined by its position in the sequence of amino acids. For example, (1) When Phe is in the next position of the start codon ATG, its meanings is that the input variable is the sensor. (2) When Tyr follows Phe, Tyr means that the number of sensors is 2. (3) In this case succeeding two amino acids determine the sensors to be used. In the same way, the meanings of amino acids in the sequence are determined by the translation rules in Table 1. Fig.8 shows an example of the DNA chromosome and genes (fuzzy rules). In this figure, bases are read from the head of the DNA chromosome, and if the start codon ATG is found, a fuzzy rule starts from this part. In this example, the next codon GCT is Alanine, and Alanine here means that the input variable is sensor. The sequence of Ala, Ser, Leu means that this sensor is No.0 sensor (input variable is D_0). The next part of Gly, Cys determines the form of membership functions, the central position x_c

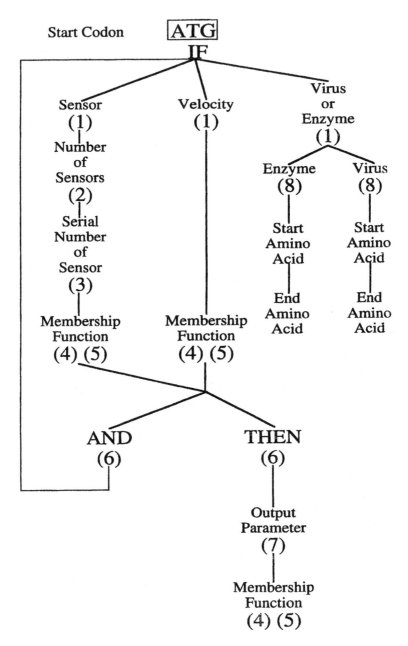

Fig.7　The Flow of Translation from the DNA Chromosome into a Fuzzy Rule

Table 1   Correspondence of Each Amino Acid and Parameter

(1)-(8) \ Amino Acid	Phe	Leu	Ile	Val	Ser	Pro	Thr	Ala	Tyr	His	Gln	Asn	Lys	Asp	Glu	Cys	Trp	Arg	Gly
(1) Input Parameter	Sensor																	*1	*2
(2) Number of Sensors	1								2				3						4
(3) Serial Number of Sensor	0	9	3	10	2	11	1	8	4	7	5		6						
(4) Central Position of Membership Fuction X	Left ←																	→	Right
(5) Width of Membership Fuction σ	Narrow ←																	→	Wide
(6) AND or THEN	AND								THEN										
(7) Output Parameter	Steering Angle u								Velocity V						u and V				
(8) Virus or Enzyme	Enzyme								Virus										

*1: Velocity   *2: Virus or Enzyme

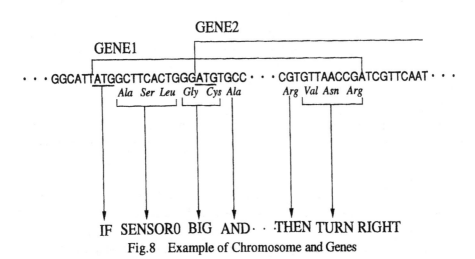

Fig.8   Example of Chromosome and Genes

and the width σ, for the input D_0. The next codon GCC makes also Alanine, and this Alanine here means AND. Like this, each amino acid has several meanings, and one meanings is selected based on the position in the gene. In this example, GENE2 in addition to GENE1 can be read from the DNA chromosome. The amino acids for virus and enzyme operations defined in this paper are explained in 4.3.

4.2 Genetic operators

There are seven chasing robots and seven avoiding robots in one generation. Each robot, either chasing or avoiding, is tested twice by randomly choosing its opponents from the seven counterparts. Each robot has initial payoffs E_1. Each test ends when either of the two confronting robots crashes into the wall or the chasing robot catches the other or a certain amount of time has passed. The payoffs for fitness values of the chromosomes are given as follows:

If the robot crashes into the wall, the robot loses payoffs E_w;

If the chasing robot reaches the avoiding robot within a fixed time, the chasing robot receives payoffs E_c, conversely the avoiding robot loses payoffs E_a;

If the chasing robot does not reach the avoiding robot in a fixed time, the chasing robot loses payoffs E_m;

If the avoiding robot reaches the foods, the avoiding robot gains payoffs E_f per a food.

After these tests are done, the genetic operations are applied to the chromosomes of the robots by regarding the payoffs of the robots as their fitness values. One chromosome of chasing robot which has the smallest payoffs is deleted. Two chromosomes which are selected from the remaining six robots are reproduced, and one-point crossover is applied to them to generate a new chromosome for a chasing robot. The chromosome of avoiding robot is newly generated in the same way. There is no constraint on the crossover points as described in section II. The mutation operator is also applied to the newly generated chromosomes. The mutation operation can be done simply by changing the bases. This mutation is done to each base at a rate of P_m. The payoffs of all robots are reset at E_1 again. After another simulation for the new generation is done, the genetic operations are applied again. These steps are repeated, and fuzzy rules which control the robots to chase or avoid the other robot and to avoid the wall are expected to be evolved.

4.3 Virus and enzyme operation

We make each chromosome contain virus and enzyme rules. The flow of translation from the DNA chromosome to these virus and enzyme rules in this paper is also shown in Fig.7. The translation rules are defined in Table 1. If one of codons GG* which make amino acid Gly is found in the DNA chromosome, the next codon decides virus or enzyme rule. The third and fourth codons determine the start amino acid and the end amino acid of virus or enzyme, respectively. If more than two rules are in one chromosome, the first virus or enzyme rule will be used. A virus is copied from other randomly selected DNA chromosome. A virus starts from the start amino acid and ends at the end amino acid determined by the virus rule and moves into the chromosome. The codon ATG must be included in between the start amino acid and the end amino acid. By this virus operation, at least one more rule is added into the DNA chromosome.

An enzyme cuts the part of the chromosome between the start amino acid and the end amino acid. The codon ATG must also be included in this part. If the payoffs of the robot becomes less in the next generation compared with that in the previous generation, the changed part by the virus or enzyme operation restored to the previous part. The virus or enzyme operation onto the part will be prohibited for the future generations. The chromosome on which the virus or enzyme operation was once applied has immunity of that virus or enzyme to prevent the same chromosome from coming into the same virus or enzyme.

V. SIMULATION

Simulations were done. The length of each initial DNA chromosome was 500. The payoffs E_l, E_w, E_c, E_a, E_m, E_f were 0, 20, 50, 20, 20, 5, respectively. Though we determined these parameters after several trials, the effects of changing these parameters are little. The probability of the mutation P_m was 0.05. The chasing robot which could not reach the avoiding robot within 30 seconds lost payoffs E_m. The genetic operations were applied to the chromosomes of chasing and avoiding robots for 100 generations alternately. If the genetic operations are applied to both simultaneously, the environment for each robot changes too rapidly to adapt.

The effectiveness of the proposed DNA coding method and the mechanism of development from the artificial DNA was examined. The following three cases were compared: (i)redundancy, overlapping, virus and enzyme operations, (ii)redundancy and overlapping, (iii)no redundancy nor overlapping. In the case of (iii), no start codon is used. The DNA is translated from its head.

Fig.9 shows the average of all payoffs of 7 chasing robots during the 3000-5000th generation. The solid line is the result of case (i), the dense dotted line is that of the case (ii), and the sparse dotted line is that of the case (iii). Each is the average of 30 trials. In a trial, the simulation from the initial generation to the 5000th generation was done. Since the genetic operations were applied to the chromosomes of chasing and avoiding robots for 100 generations alternately, the payoffs fluctuated periodically. The redundancy, overlapping, and virus and enzyme operations worked well. Fig.10 shows the average of the payoffs of 7 chasing robots of another simulations. The solid line is the case (iv) where the chasing robots used the proposed coding method and the avoiding robots used the conventional method in [12]. The dotted line is the case (v) where the chasing robots used the conventional method and the avoiding robots used the proposed coding method. In both the cases, no virus nor enzyme operations were used. Each is the average of 10 trials. The parameters of the genetic operations of the conventional method were tuned and the result in Fig.10 was the best among we obtained. This figure shows the advantage of the proposed method having the redundancy and overlapping of genes. An example of evolution of control rules of chasing and avoiding robots is shown in Fig.11. Fig.11(a) shows the initial position and direction of the chasing robot and the avoiding robot. Fig.11(b), (c), (d), (e) and (f) shows an example of each movement at the 300th, 400th, 500th, 2300th, and 5000th generation, respectively. In Fig.11(b), the avoiding robot turned right at first and went straight. The chasing robot only went straight and reached the avoiding robot. In Fig.11(c), the avoiding robot went straight and the chasing robot crashed into the wall. In Fig.11(d), the chasing robot turned right at first and went straight and reached the avoiding robot. In Fig.11(e), each movement was became a little complex. The chasing robot moved around in the upper left area, and the avoiding robot drew a rhombus. For the avoiding robot, the rhombic movement was able to get many food, but it was dangerous to be caught by the chasing robot. Fig.11(f) was the case where the avoiding robot acquired very effective rules. The avoiding robot was running along the wall. The chasing robot could measure only the distance using the ultra-sonic sensors. Therefore, the avoiding robot had, so to speak, protective coloring. The avoiding robot could avoid the chasing robot for 30 seconds. The followings are the examples of acquired rules of the avoiding robot. These were the rules mainly used to run along the walls. The membership functions determined by the central position x_c and the width σ in the DNA chromosome are labeled for clarity by the authors.

IF D_0 is Medium AND D_2 is Medium AND D_1 is Far AND D_{11} is Medium
    THEN u is Very Small Left, V is Medium Small

IF D_1 is Medium AND D_0 is Medium AND D_{11} is Medium
    THEN u is ZERO, V is Medium Big

IF D_3 is Near AND $D0$ is Far THEN u is Small Right

IF D_0 is Near THEN u is Medium Right

These rules are used to go along the walls on the left-hand side. In this case, the avoiding robot acquired very effective rules. In other cases, the chasing robot acquired effective rules. For example, the chasing robots ran along the walls looking for the avoiding robots.

VI.CONCLUSIONS

This paper proposed a new coding method, DNA coding method, and the mechanism of development from the artificial DNA. This DNA coding method is suitable for knowledge representation. The DNA chromosome has a redundancy, and allows overlap of genes. This paper applied the proposed DNA coding method to knowledge acquisition of fuzzy rules based on chasing and avoiding actions of mobile robots. The new coding method was efficient in finding fuzzy rules. The proposed method also made applications of virus and enzyme operations easy.

PAYOFF

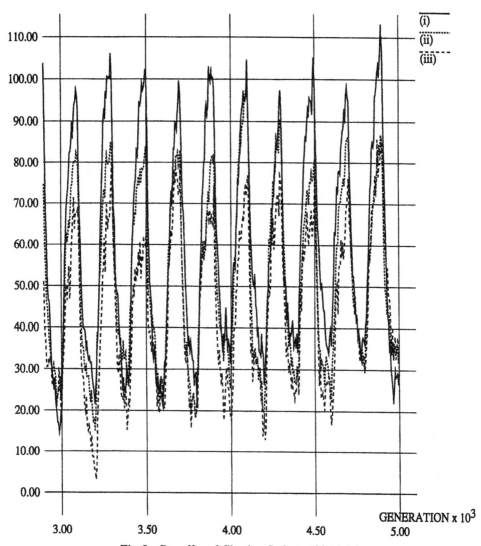

Fig.9   Payoffs of Chasing Robots (30 trials)

PAYOFF

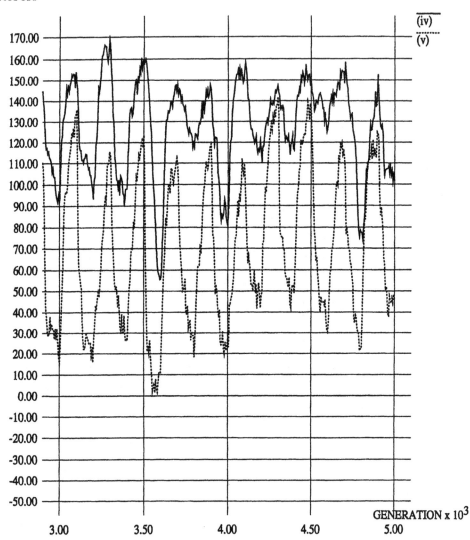

GENERATION x 10^3

Fig.10   Payoffs of Chasing Robots (10 trials)

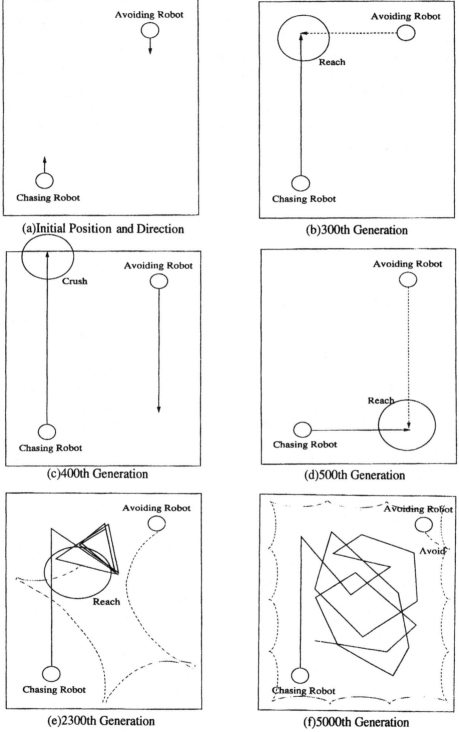

(a)Initial Position and Direction

(b)300th Generation

(c)400th Generation

(d)500th Generation

(e)2300th Generation

(f)5000th Generation

Fig.11   The Movement of Each Robots

REFERENCES

[1]D.E.Goldberg, *"Genetic Algorithm in Search"*, Optimization and Machine Learning, Addison Wesley (1989)

[2]L.Davis(Editor), *"Handbook of Genetic Algorithm"*, Van Nostrand Reynold (1989)

[3]W.Wienholt, *"A Refined Genetic Algorithm for Parameter Optimization Problems"*, Proceedings of The Fifth International Conference on Genetic Algorithms (1993)

[4]C.L.Karr, L.Freeman, D.Meredith, *"Improved Fuzzy Process Control of Spacecraft Autonomous Rendezvous Using a Genetic Algorithm"*, SPIE Conference on Intelligent Control and Adaptive Systems, pp.274-283 (1989)

[5]C.L.Karr, *"Design of an Adaptive Fuzzy Logic Controller Using a Genetic Algorithm"*, Proceedings of the 4th International Conference on Genetic Algorithms, pp.450-457 (1991)

[6]M.Valenzuela-Rendon, *"The Fuzzy Classifier System: A Classifier System for Continuously Varying Variables"*, Proceedings of the 4th International Conference on Genetic Algorithms, pp.346-353 (1991)

[7]J.H.Holland, J.S.Reitman, *"Cognitive Systems Based on Adaptive Algorithms"*, in Pattern Directed Inference Systems, D.A.Waterman, F.HayesRoth (Editors), pp.313-329. Academic Press, New York (1978)

[8]T.Furuhashi, K.Nakaoka, K.Morikawa, Y.Uchikawa, *"Controlling Excessive Fuzziness in a Fuzzy Classifier System"*, Proceedings of the 5th International Conference on Genetic Algorithms, p635 (1993)

[9]T.Furuhashi, K.Nakaoka, K.Morikawa, Y.Uchikawa, *"An Acquisition of Control Knowledge Using Multiple Fuzzy Classifier Systems"*, Journal of Japan Society for Fuzzy Theory and Systems, Vol.6, No.3, pp.603-609 (1994)

[10]K.Nakaoka, T.Furuhashi, Y.Uchikawa, *"A Study on Apportionment of Credits of Fuzzy Classifier Systems for Knowledge Acquisition of Large Scale Systems"*, Proceedings of the 3rd International Conference on Fuzzy Systems, pp.1797-1800 (1994)

[11]M.A.Lee, H.Takagi, *"Dynamic Control of Genetic Algorithms Using Fuzzy Logic Techniques"*, Proceedings of the 5th International Conference on Genetic Algorithms, pp.76-83 (1993)

[12]T.Hashiyama, T.Furuhashi, Y.Uchikawa, *"A Study on Fuzzy Rules for Semi-Active Suspension Controllers with Genetic Algorithm"*, Proceedings of IEEE International Conference on Evolutionary Computation (ICEC'95), PP.279-282, (1995)

[13]T.Furuhashi, Y.Miyata, K.Nakaoka, Y.Uchikawa, *"A New Approach to Genetic Based Machine Learning and an Efficient Finding of Fuzzy Rules"*, Lecture Notes in Artificial Intelligence, Vol.1011, pp.173-189 (1995)

[14]B.Albers and others, *"Molecular Biology of the Cell"*, Garland Publishing (1994)

[15]A.Kornberg, *"DNA Synthesis"*, W.H.Freeman and Company (1974)

Experimental Study on Acquisition of Optimal Action for Autonomous Mobile Robot to Avoid Moving Multiobstacles

*Takeshi AOKI,   **Toshiaki OKA,   ***Soichiro HAYAKAWA,
**Tatsuya SUZUKI,   **Shigeru OKUMA*

*Nagoya Municipal Industrial Research Institute
   3-4-41, 6-ban, Atsuta-ku, Nagoya, 456,   JAPAN
   TEL&FAX: +81-52-654-9932, Email: aoki@okuma.nuee.nagoya-u.ac.jp
**School of Electric Engineering, Nagoya University
   Furo-cho, Chikusa-ku, Nagoya, 464-01,   JAPAN
   TEL: +81-52-789-2778, FAX: +81-52-789-3140
   Email:oka,suzuki,okuma@okuma.nuee.nagoya-u.ac.jp
***Toyota Institute of Technology
   2-12, Hisakata, Tenpaku-ku, Nagoya, 468,   JAPAN
   Tel:+81-52-809-1827, Fax:+81-52-809-1721; Email: s-hayakawa@toyota-ti.ac.jp

Abstract

The principal aim of this study is to show how an autonomous mobile robot can acquire the optimal action to avoid moving multiobstacles through the interaction with the real world. In this paper, we propose a new architecture using the hierarchical fuzzy rules, fuzzy performance evaluation system and learning automata. By using our proposed method, the robot acquires the fine behavior to move to the goal, avoiding moving obstacles, simultaneously by using the steering and velocity control inputs. Also we show the experimental results to confirm the feasibility of our method.

1. Introduction

Many kinds of methods for an autonomous mobile robot to avoid obstacles using fuzzy logic have been studied[1-3]. Most of them, however, adopt an only steering control. In the previous researches, methods which adopt both steering and velocity controls have hardly been found[8][9], nevertheless its potential of achieving high performance[4]. This is because the number of fuzzy rules increases in proportion to the square number of control inputs and also because adjusting and modifying fuzzy rules come to be difficult. To overcome this problem, more simple fuzzy logic structure for multicontrol inputs system should be considered. In the precious paper[5], we have proposed an three-stage algorithm for obstacles avoidance using hierarchical fuzzy rules. At the first stage, the robot calculates and decides the dangerous obstacles among the ones. Then, using the hierarchical fuzzy rules, the autonomous mobile robot calculates two control inputs for each obstacle avoidance, that is, steering and velocity control inputs, respectively. At the next stage, the robot adjusts each steering control input using the fuzzy balancer in order not to conflict with the velocity control input for each obstacle. At the final stage, the robot arbitrates and combines all the steering control inputs with weights for moving to the goal avoiding the obstacles without degrading the control performance.

However the drawback of this method are the difficulties of the arbitration between these two control inputs and combination among steering control inputs. In other words, it is difficult for a designer to make the adjustment and combination rules in advance. This is because the designer can not get all information about the real world and image all situations the robot will face in advance. As the results, many trials are needed to design the rules. To overcome this problem, the robot should introduce a learning system. Especially, we focus on how an mobile robot can autonomously acquire the optimal action through the interaction with the real world.

In this paper, we propose a new action selection technique based on the framework of learning automata[6-7]. Our proposed system consists of 6 modules: *Sensor Module, Hierarchical Fuzzy Rule Module, Action Generating SubModule and MetaModule, Fuzzy Performance Evaluation Module, and Learning Automaton Module*. In the learning automata, each adjustment or combination rule is regarded as one action, and several actions are prepared in advance. Then, using this learning method in each action generating module, the robot can autonomously acquire the optimal rule suitable for each situation through the interaction with the real world. Our proposed system has the flexible structure to adapt the dynamical world because basic control inputs are decided based on the fuzzy logic in the hierarchical fussy rule module and fine adjustment or combination of the control inputs are acquired by the reinforcement learning algorithm in the action generating modules. The reinforcement learning method on the framework of the learning automata used in this study has the property called "absolute expediency", which means that action selection probabilities are guaranteed mathematically to converge under the unknown static random environment. Moreover, This learning algorithm is executable in real time because of its simplicity.

In case of using learning methods like the reinforce learning ones, we also have to consider how the performance is evaluated. The world always changes and the performance index is inclined to be complicated. In our proposed system, we construct the evaluating function using the fuzzy logic. This is because the fuzzy logic can reflect the designer's experiences or intuitions explicitly and evaluate the performance easily.

Using our proposed system, the robot performs the learning cycle "sense-select-act-evaluate-learn(update)" in the real world. Finally, the robot autonomously acquires the optimal action selection suitable for the real world. In the following section, the details of our system are explained.

2. Proposed System for Moving Obstacle Avoidance

2.1 Architecture for Single Obstacle

In this subsection, we state the outline of our proposed system. At first, the system for single obstacle is explained. Then, the extended system for multiple obstacles is addressed. The proposed architecture for single obstacle is shown in Figure 1. The system consists of 5 modules. Each module is as follows.

1)*Sensing Module*: This module senses the environment around the robot and calculate the relative positions and velocity vectors between the robot and obstacle.

2)*Hierarchical Fuzzy Rule Module*: This module decides the steering and velocity control inputs for obstacle avoidance, respectively.

3)*Action Generating SubModule*: This module selects one action (rule) among actions based on the action probabilities explained later, in order to combine the steering control input with the velocity control input based on the selected action.

4)*Fuzzy Performance Evaluation Module*: This module evaluates how good or bad the selected action (rule) is by using the fuzzy logic.

5)*Action Learning Module*: This module updates and memorize the action probabilities in the element of the table based on the evaluation value from the evaluation module by the reinforcement learning method in Learning Automata.

One of several actions, which correspond to the combination rule, is selected according to the action probabilities in the element of the probability table. The environment around the robot is mapped into this table. Note that each element of the table reveals the situation that the robot faces. The details on this table will be explained in the later section. In the following sections, the detail of each module is described.

2.1.1 Hierarchical Fuzzy Rules

The advantage of this hierarchical algorithm is that the number of fuzzy rules can be reduced by grouping fuzzy rules as a module using specific medium parameters and piling fuzzy logic modules up. By using the proposed hierarchical algorithm, the more the number of input parameters increases, the bigger the effectiveness of reducing rules comes out. Moreover, the property and the feature of the algorithm are its reconstructability owing to the modular construction. Figure 2 shows the structure of hierarchical fuzzy rules, the definition of the fuzzy parameters and the relations between inputs and outputs. The decision steps in this algorithm are as follows.

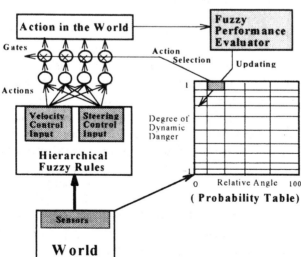

Figure 1.   Proposed System for Single Obstacle

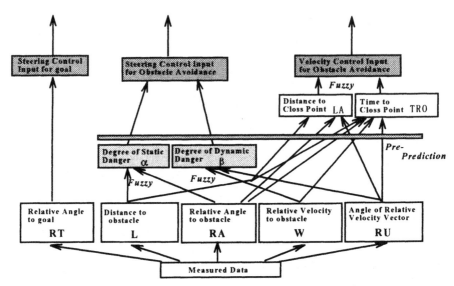

Figure 2.   Hierarchical Fuzzy Rules

1)The measured data of the position and velocity vectors of the robot, the obstacle and the goal are input to the algorithm.

2)The state parameters are calculated based on these data. The state parameters are,

-**RT:** the relative angle between moving direction of the robot and the goal direction,

- **L :** the distance between the robot and the obstacle,

-**W:** the relative velocity between the robot and the obstacle ,

-**RA:**the relative angle between moving direction of the robot and the direction from the robot to the obstacle,

-**RU:** the relative velocity angle between W and RA

3)The degree of static danger α[5] is a fuzzy parameter which is calculated from the fuzzy rules with the relative angle RA and the distance L. The value of α is restricted within ± 1. The degree of dynamic danger β[5] is also a fuzzy parameter which is calculated from fuzzy logic with the relative angle RU and the velocity W. It is also restricted within ± 1. The parameter α indicates the degree how dangerous the moving obstacle is in the static relation between the robot and the obstacle. The parameter β indicates the degree how dangerous the moving obstacle is in the dynamic relation between them. For example, α=-1 indicates the situation that an obstacle is located in the front and left side of the robot and very close. β=-1 indicates the situation that an obstacle approaches the robot fast and will pass it on the left side. The details of these parameters are explained in [1].

4)The steering control input is decided by the steering table[5] using α and β. The velocity control input is decided by the fuzzy rules[5] using the two medium parameters shown in Figure 2. These details of this structure is shown in [5].

2.1.2 Actions and Probability table of Action Generating Submodule for Single Obstacle

As shown in Table 1, the probability table of the submodule can be considered as the two-dimensional finite space of which dimensions represent the dynamic danger β and the relative angle RA. The environment around the robot is mapped into this table. In other words, each element of the table reveals the situation that the robot faces. The parameter β as described before indicates the degree how dangerous the moving obstacle is in the dynamic relation between the robot and the obstacle. If the value of β is 1, it means that the obstacle is approaching the robot straight with relative velocity W. And when the value of β is 1 and the relative angle RA is 90°, the situation is that an obstacle is approaching the robot from the side. When the value of β is 1 and the relative angle RA is 0°, the obstacle is approaching the robot from the front. The robot judges the situation and decides the element in the table using the value of β and RA.

Each action in Figure 1 corresponds to the combination rule of the steering control input so as to support the velocity control input. There are four actions in our system, and the role of them is listed in the following.

Action 0: To use both the steering and velocity control inputs for goal,

Action 1: To use the steering control input for goal and the velocity control input for obstacle avoidance,

Action 2: To combine half of the steering control input for obstacle avoidance with half of it for goal, and use the velocity control input for obstacle avoidance,

Action 3: To use both the steering and velocity control inputs for obstacle avoidance,

Where the velocity control input for goal of Action 0 means that the robot dose not control its velocity to avoid a moving obstacle. The key idea of designing these actions is that the velocity control input has priority over the steering control input for obstacle avoidance. This is because the velocity control is decided using geometrical information and reflect the actual situation appropriately in our proposed algorithm [5]. Each action has its own probability in any situation, that is, in every element of the probability table. By the roulette method, the robot selects one action among four based on the action probabilities in the element of the table, which reveals the corresponding situation that the robot faces in the world. The robot acts in the world using the following equation.

$$Sc = (1 - w) \cdot Sg + w \cdot Sa$$

where   Sc : Final Steering Control Input   Sg : Steering Control Input for Goal
       Sa : Steering Control Input for Obstacle Avoidance
       w: weight based on Selected Action

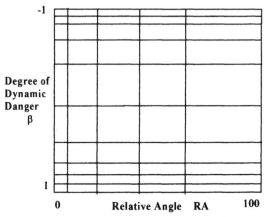

Table 1.   Probability Table for Single Obstacle

2.1.3 Learning Scheme

It is difficult for a designer to specify the selection rule of the action in advance because the designer can not get all the information about the real world in advance. One of the most promising method to overcome this problem is adopting the algorithm based on the framework of Learning automata[6]. Using this reinforcement learning method, the robot can autonomously acquire the good action in any situation through the interaction with the real world. As the reinforcement scheme, we use the PLr-i algorithm[6]. The PLr-i method updates the action probabilities based on the binary value of evaluation. When the robot action is good, that is, the output from the fuzzy performance evaluator described in the later subsection is positive, the action probabilities are updated as follows.

PLr - i Scheme

$$\begin{cases} p_i(n+1) = p_i(n) + c\big[1 - p_i(n)\big] & \text{if } a(n) = a_i \text{ and } b(n) = 1 \\ p_j(n+1) = (1-c)p_j(n) & \text{for } j \neq i \end{cases}$$

$$p_i(n+1) = p_i(n) \qquad \text{for all } i, \text{ if } b(n) = 0$$

where c is learning parameter, a_i and $a(n)$ denote the ith action and selected action at the nth stage, respectively. $P_i(n)$ denotes the probability of action i at the nth stage and $b(n)$ is the output of the fuzzy evaluation system at the nth stage. When the robot action is bad, that is, the output from the evaluator is not positive, the action probabilities are not changed.

Each element of the probability table has four probabilities which correspond to four actions. The initial value of each probability is 1/4. After choosing an action according to the probabilities of the element which reveals the situation of the robot, the robot updates the probabilities of the element with the evaluation value using PLr-i method. The main idea of the PLr-i algorithms is that the absolute expediency in the unknown random world is guaranteed[6]. This is due to the operation of "inaction" in case that the evaluation is not good.

2.1.4 Fuzzy Performance Evaluator for Single Obstacle

In case of using the reinforcement learning method explained in the previous subsection, one of the difficulties is how to evaluate the performance. Generally speaking, it is difficult to find standards for evaluation and calculate the degree how good or bad the action is. Moreover, the world is always changing.

It is why the performance index is inclined to be complicated. One of the natural idea to treat this problem is to use the fuzzy logic. By using fuzzy logic, we can reflect our experiences or intuitions to the performance evaluation. In our proposed system, the fuzzy logic is constructed from 2 inputs and 1 output. The fuzzy input parameters at the nth stage are,

1) The difference between the degree of dynamic danger at the (n-1)th stage and the one at the nth stage.
2) The difference between the distance at the nth stage and the one at the (n-1)th stage from the robot to the goal.

Also the fuzzy output parameter at the nth stage is,

3) the evaluation value between +1(GOOD) and -1(BAD)

The fuzzy evaluation rules are shown in Table 2. Using these tables, the fuzzy evaluation system calculates the degree how good or bad the robot action is.

Table 2. Evaluation Fuzzy Rules for Single Obstacle

		Parameter ② Change of Distance Between Robot and Goal		
		Negative	Zero	Positive
Parameter ① $\lvert \beta(n) \rvert - \lvert \beta(n\text{-}1) \rvert$	Negative Big	Positive Big	Positive Medium	Positive Medium
	Negative Small	Positive Small	Positive Small	Zero
	Positive Small	Negative Small	Negative Small	Negative Medium
	Positive Big	Negative Medium	Negative Big	Negative Big

2.2 Actions and Probability table of Action Generating Metamodule for Multiple Obstacles

2.2.1 Additional Module

Here, We present a new extended system for avoiding multiple obstacles as shown in Figure 3. This proposed system consists of 6 modules. This extended system is that the Action Generating MetaModule is added to the system for single obstacle. This additional module also selects one action (rule) among actions based on the action probabilities in order to combine the adjusted steering control inputs for each obstacle with weights based on the selected action. This module has 4 actions which means the number of obstacles to be paid attention to. Note that the Action Generating SubModule is assumed to be learned enough in advance. Then four combination rules are prepared as actions like below.

Action 0: To ignore obstacles and use the steering control input for goal,
Action 1: To use the only steering control input from the submodule of the most dangerous obstacle in the sense of dynamic danger.
Action 2: To use the steering control inputs from the submodules of the most and secondary dangerous obstacle in the sense of dynamic danger.
Action 3: To use all the steering control inputs from submodules.

In other words, the action reveal the number of obstacles the robot should pay attention to. Each action also has its own probability in any situation, that is, in every element of the probability table of the metamodule shown in Table 3. By the roulette method, the robot selects one action among four in the element of the table explained later. According to the selected action, robot calculates steering control input like below.

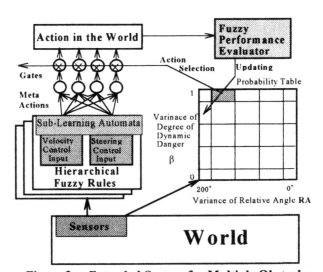

Figure 3. Extended System for Multiple Obstacles

Selected Action No. $r = 0$: $Sc = Sgl$

Selected Action No. $r > 1$: $Sav = \dfrac{\sum\limits_{i=1}^{r} \beta i \cdot Si}{\sum\limits_{i=1}^{r} \beta i}$, $Sc = (1 - \beta \max) \cdot Sgl + \beta \max \cdot Sav$

Where r denotes the selected action, βi as the degree of dynamic danger for Obstacle i, $\beta \max$ as the maximum value among βi, Si as the adjusted steering control input from submodule for Obstacle i, Sgl as the steering control input for goal and Sc as the combined steering control input. The robot controls velocity using the control input of most dangerous obstacle in the sense of dynamic danger. Then, the robot performs the selected action in the world with both control inputs.

The probability table of the metamodule can be considered as the two-dimensional finite space of which dimensions represent the variance of the dynamic danger β and the variance of the relative angle RA. The variance of the parameter β indicates a kind of the distribution of the dangerous moving obstacle. If the value is 0, it means that all the obstacle are equally dangerous in the sense of dynamic danger. And when the value is 1, it means that one obstacle approaches the robot directly and one is not dangerous for the robot. On the other hand, when the variance of the relative angle RA is 0, it means that the obstacles are located at the same direction. If the variance is 200, it means that one obstacle is located at the left side of which angle is -100 degree and one is located at the right side of which angle is 100 degree.

2.2.2 Fuzzy Performance Evaluator for Multiple Obstacles

In the extended system, the fuzzy logic is also used to evaluate the selected action from the metamodule and constructed from 2 inputs and 1 output. The fuzzy input parameters at the nth stage are,

1) The difference of the maximum value of the degree of dynamic danger between at the $(n-1)$th stage and at the nth stage.

Table 3. Probability Table of MetaModule
Variance of Relative Angle   RA

2) The difference between the distance at the nth stage and the one at the $(n-1)$th stage from the robot to the goal.

Also the fuzzy output parameter at the nth stage is,

3) the evaluation value between +1(GOOD) and -1(BAD)

The fuzzy evaluation rules are shown in Table 4. Using these tables, the fuzzy evaluation system calculates the degree how good or bad the robot action is.

3. Experimental Study

In this section, the experimental results are shown to confirm the feasibility of our proposed system.

3.1 Experimental Setup

The assumptions made in the experiments are as follows.

Obstacle

1)The number of obstacles is three.

2)The start points are set at random in the front side of the robot at every run.

3)The moving directions are set so as to approach the robot

4)The Obstacles move straight with constant velocity. The velocities are set at random up to the initial velocity of the robot at every run.

Robot.

5) The start point, moving direction and initial velocity is same at every run.

6)The action probabilities are updated at each time when the robot performs one action.

Table 4. Evaluation Fuzzy Rules for Multiple Obstacles

		Parameter ②	Change of Distance Between Robot	and Goal
		Negative	Zero	Positive
	Negative Big	Positive Big	Positive Medium	Positive Medium
Parameter ①	Negative Small	Positive Small	Positive Small	Zero
$\|B(n)\| - \|B(n-1)\|$	Positive Small	Negative Small	Negative Small	Negative Medium
$B(n)=\max(\beta_1(n))$ $i=1,2,3$	Positive Big	Negative Medium	Negative Big	Negative Big

7)PLr-i algorithm is used as the learning method.

8)The learning parameter is 0.1

The mobile robot and three obstacles used in this experiment are Khepera made by EPFL. The robot is controlled by a NEC PC9801BA Computer (CPU:Intel 486DX2 40MHz) via a RS232C line by means of a cable. This cable is also used to supply electrical power. As for the sensor, CCD camera is used to get the information of the environment around the robot. The camera is set at the height of 2m from the ground in order to watch 2.5x3m area. The camera is also connected with the computer. All processes such as picture reading and analyzing, decision making and motor control are performed with sampling rate 0.2 sec by the computer. The experimental system is depicted in Figure 4. Learning of the probability tables are performed by simulation in advance. Then using those tables, the fesibility of our method is tested with real robots.

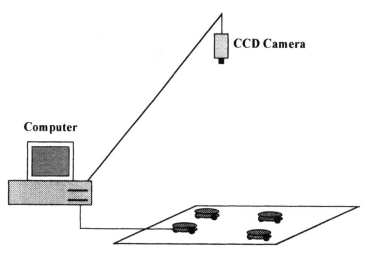

Figure 4   Experimental Setup

3.2 Experimental Results for Multiple Obstacles

Figure 5-7 show the motion histories of the robot and obstacles. In Figure 5, the robot starts learning and the action probabilities in the Action Generating MetaModule are not well-tuned. In this case, the robot often chooses irrelevant actions and bumps into the obstacles.

After learning, the robot has acquired the intelligence of action selection based on the action probabilities. In Figure 6, the robot chooses fine actions and inhibits unnecessary actions and the robot can move to the goal avoiding moving obstacles in spite of the same situation as Figure 5. Figure '7 shows the other case after learing. It also shows the robot has acquired fine behaviors on the action probability table.

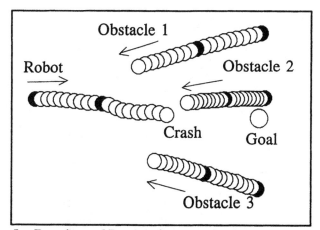

Figure 5.   Experimental Result before Learning for Multiple Obstacles

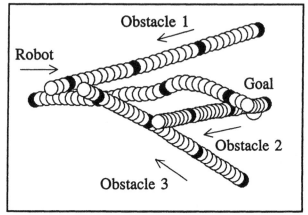

Figure 6.   Experimental Result after Learning for Multiple Obstacles

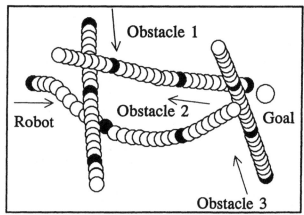

Figure 7.   Experimental Result after Learning for Multiple Obstacles

Table 5. Probability Table of MetaModule

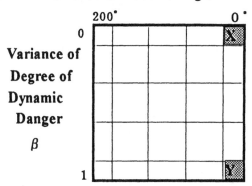

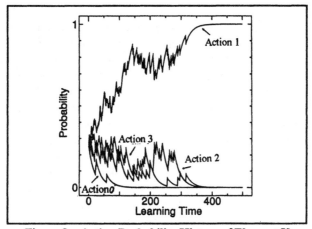

Figure 8   Action Probability History of Element X

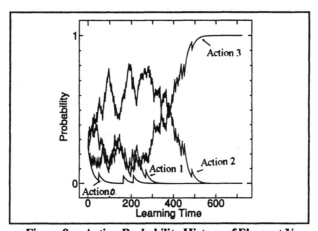

Figure 9.   Action Probability History of Element Y

Figure 8 shows the action probability histories of the element X of the probability table in Table 5. This figure indicates that the robot has acquired action rule that robot regards obstacles as one group and pays attention to just the most dangerous obstacle in the situation corresponding to element X that the obstacles are equally dangerous and located at same direction. Figure 9 also shows the action probability histories of the element Y in Table 5. This figure indicates that the robot has acquired action rule that the robot pays attention to all the obstacles because of different degrees of dynamic danger in the situation corresponding to the element Y.

4.Discussion

The hierarchical fuzzy rules and the Action Generating SubModule are applied to each obstacle. The number of obstacles, which the robot should pay attention to, has an influence not on the hierarchical fuzzy rules but on the Action Generating MetaModule. Therefore, the amount of calculation dose not increase so much even if the number of obstacles increases compared with that of this study.

It is difficult for the designer to make all the adjustment and combination rules of the control inputs by taking account for all the situation the robot faces in advance, in order not to conflict one another and degrade the control performance. To overcome this problem, we use the reinforcement learning method on the framework of the learning automata and let the robot acquire the optimal action rules automatically through the interaction with the real world. The learning experiences are memorized in the action probability table. In each element corresponding to each situation the robot faces, the optimal action probability has the highest value or converge to 1. But there are some elements where the action probabilities always change and dose not converge. In those cases, it means that multiple action rules are fine for the situation and it leads to the variation of actions.

However, note that it is difficult for the robot to have enough experiences for each element , that is, situation uniformly. Generally speaking, the learning convergence rate depends on times of facing situations. If the robot faces a situation very often, the action probabilities of the element converge rapidly. But if a situation dose not occur so often, the probabilities are hardly updated. This results in different convergence speed in the different situation. To overcome this problem, the different learning parameters need be used in each element of the probability table. For example, in the element the robot faces very often, the small learning parameter is used. In the element the robot faces not so often, the large parameter need be used.

5.Conclusion

We propose a new action selection technique for moving multiobstacles avoidance using hierarchical fuzzy rules, fuzzy evaluation system and learning automata through the interaction with the real world. By using our proposed system, an mobile robot autonomously acquires the fine behaviors how to move to the goal avoiding moving multiobstacles using the steering and velocity control inputs, simultaneously. At the first learning step, our proposed system learns and acquires the fine behaviors for single moving obstacle. Then the system completes the action probability table of the action generating submodule and memorize fine behaviors. At the second learning step, the system learns and acquires the optimal actions for avoiding multiple moving obstacles on the action generating metamodule. We show the feasibility of our proposed system with experimental results.

References

[1]Y.Maeda et.al,"Collision Avoidance Control among Moving Obstacles for a Mobile Robot on the Fuzzy Reasoning," J.of the Robotics Society of Japan, Vol6, No.6, pp.518-522, 1988.

[2]H.Koyama et.al,"Study of Obstacle Avoidance Problem for Mobile Robot Using Fuzzy Production System," J.of the Robotics Society of Japan, Vol9, No.1, pp.75-78, 1991.

[3]T.Takeuchi,"An Autonomous Fuzzy Mobile Robot," J.of the Robotics Society of Japan, Vol6, No.6, pp.549-556, 1988.

[4]S.Ishikawa,"A Method of Autonomous Mobile Robot Navigation by Using Fuzzy Control," J.of the Robotics Society of Japan, Vol9, No.2, pp.149-161, 1991

[5]T.Aoki et.al,"Motion planning for Multiple Obstacles Avoidance of Autonomous Mobile Robot Using Hierarchical Fuzzy Rules," 1994 IEEE International Conference on Multisensor Fusion and Integration for Intelligent Systems, pp.265-271.

[6]K.Narendra et.al.,"LEARNING AUTOMATA", Prentice-Hall. Inc.,1989

[7]M.F.Norman.,"Markov Processes and Learning Models", Academic-Press.,1972

[8]H.NAGATA et.al.,"Intelligence Control Concerning Obstacle Avoidance of Mobile Robot", J. of the Robotics Society of Japan, Vol.11, No.8, pp.1203-1211, 1993

[9]M.ITO et.al.,"Collision Avoidance among Autonomous Vehicles Containing Fixed Obstacles", Proc. of the 35th Annual Conference of the Institute of Systems, Control and Information Engineers, May 22-24, pp.63-64, 1991

New Approaches on Structure Identification of Fuzzy Models:

Case Study in an Electro-Mechanical System

P.J. Costa Branco[*], N. Lori[**] and J.A. Dente[*]

[*] Instituto Superior Técnico
CAUTL/Laboratório de Mecatrónica
Av. Rovisco Pais, 1096 Lisboa Codex, Portugal
Fax Nº. 351-1-8417167
E-mail: pbranco@alfa.ist.utl.pt
http://macdente.ist.utl.pt

[**] Washington University in Saint Louis
Physics Department
USA
E-mail: lori@hbar.wustl.edu

The main problem in design fuzzy models is to identify their structure. This means recognise the variables that better characterise the system dynamics, the number of membership functions partitioning each variable, as well as their distribution and fuzziness degree. This work presents two pre-processing methods for structure identification of fuzzy models. The first approach uses the statistical method of Principal Component Analysis (PCA). The second one uses a clustering technique called *autonomous mountain-clustering method*. The statistical method of Principal Component Analysis helps to select the variables that dominate the system dynamics. Besides, this method contributes to design fuzzy models with better performance. The second approach identifies the fuzzy model order. That is, the method identifies the number of membership functions attributed to each variable, as well as their position and width. So, the *autonomous mountain-clustering* eliminates the usual "trial-and-error" mechanism. The pre-processing methods can be used to initialize the neuro-fuzzy techniques and therefore accelerate their learning process. We test these methods using a simple learning process applied to extract the fuzzy model of an experimental electro-hydraulic system. The results show that a good modeling capability is achieved without employ any complicated optimisation procedure to structure identification.

1 Introduction

The structure of a fuzzy model can be extracted from expert's knowledge by translating their information about the system to a linguistic description. An example is the knowledge representation of the human operator decisions in process control. Clearly, an identification of the system's structure based only in the expert's description can be very poor. If the acquired information is wrong or not enough, the model will be bad. It is necessary to complement operator subjectivity with more objective knowledge using available numerical data from the system in question.

Structure identification of fuzzy systems [6] consists in detect the variables that dominates the system dynamics, the identification of the number of membership functions partitioning each variable, their position in the respective universe of discourse, as well as their fuzziness degree indicated by each fuzzy set's width.

In dynamical systems the number of variables is often very high. If the premise part of each fuzzy rule considers all variables, the model will have great complexity with high computational costs. However, in real situations only few variables dominate the process dynamics. The statistical technique of PCA helps to define these set of variables reducing the effective dimensionality of the system.

In learning procedures, the working domain is not completely filled by the examples. Under this situation, the fuzzy model extracts false rules localised in the empty regions. For these rules, the inference process fails originating large errors. We use the PCA analysis defines a new variable set denoted by principal components. This new working domain becomes less sparse and the available information is expanded increasing the fuzzy model performance.

In fuzzy modeling and fuzzy control, the number of membership functions assigned to each variable and their position in the universe of discourse are responsible for the acquired performance. The respective position and number of membership functions reflect the distribution of learning examples in the working domain. A clustering technique developed by the authors, called *autonomous mountain-clustering method* [2], identifies their number, position, and fuzziness degree.

The algorithm provides us with a set of clusters describing the data set structure. Each cluster, projected onto the coordinate axes, results in a membership function situated in the respective cluster coordinate. The clustering algorithm eliminates the "trial-and-error" process usually used to compute the fuzzy partition of the working domain. It decreases the fuzzy model complexity because reduces the number of rules which describe the system, and it automatically extracts from the learning set the membership function parameters without employ any complicated optimisation procedure.

Fuzzy logic becomes a new useful mean for automatic modeling electro-mechanical systems [3,4]. Simple applications for these model-learning algorithms are the compensation of nonlinear terms that affect system dynamics and their implementation in "intelligent" systems with self-learning capabilities.

We employ an experimental electro-hydraulic system to investigate the potentialities of the proposed approaches on help modeling electro-mechanical systems with fuzzy logic. A learning set, containing examples of system behaviour, is constructed using all system accessible variables. The PCA method indicates the variables that better characterise the system dynamics. We propose a modified fuzzy model that uses the new variables obtained by PCA, and we present its great potential in increasing the accuracy of the computed fuzzy model. The *autonomous mountain-clustering algorithm* is applied to the learning set and identifies the number of membership functions attributed to each variable, as well as their position and width.

The pre-processing methods can be used to initialize the neuro-fuzzy techniques. The methods can accelerate the learning process and reduce the design complexity of the neuro-fuzzy model.

The organisation of the paper is as follows. Section 2 describes the experimental system and how the learning sets are obtained. Section 3 resumes the main steps of the learning process used to extract the model. This section point out the modelling problems to justify the use of pre-processing methods. In section 4 we present the technique of PCA and its application in select the variable set dominating the system dynamics. We propose a modified fuzzy model using the principal components making more robust fuzzy models. Section 5 describes the second pre-processing method the *autonomous mountain-clustering algorithm*. This section uses the clustering algorithm to identify the premisse structure of the fuzzy models without any "trial-and-error" process. At last, section 6 presents our conclusions.

2 The Electro-Mechanical System

Fuzzy modeling of electro-mechanical systems becomes a new tool for compensation of nonlinear terms affecting systems behaviour. Another direction is implement electro-mechanical systems with self-learning capabilities to construct more autonomous machines.

Almost all control systems employ linear or locally-linearized models of the process to be controlled. If we have a more complete representation of system dynamics using fuzzy models, a better system control is possible. Besides that, the identification of the fuzzy model structure is an important procedure to decrease the computational time in find the optimal model. It reduces the model complexity because the structure identification process can identify the membership function set that better characterises the model structure.

The electro-mechanical system is composed by an electrical drive connected to a fixed displacement pump that controls the position of a hydraulic piston. The electrical drive is a commercial one, controlling the speed of a permanent-magnet synchronous motor by a proportional-integrator (PI) controller. Figure 1a shows a schematic representation of the system with its elements identified by the respective numbers. In Figure 1b, we show a picture of the experimental system indicating the main elements.

1- Voltage source inverter using IGBT's;

2- Permanent magnet synchronous motor (220V/1.2Nm);

3- Fixed displacement pump;

4- Hydraulic actuator (total displacement = 0.2m);

5- Inertial load applied to the piston;

6- Angular speed motor sensor (±3000 rpm);

7- Pressure transducer (*PI*) and the magnetostrictive linear displacement sensor for piston speed (*v*) and position (*y*) signals;

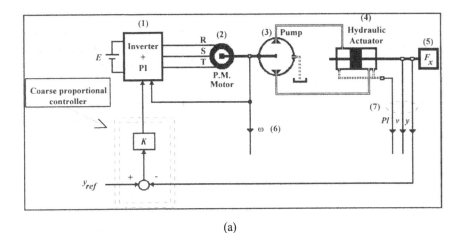

(a)

(b)

Fig. 1. (a) Schematic representation and (b) picture of the electro-hydraulic system.

The variables considered to characterise the system dynamics are: y_{ref} the reference signal for piston position; ω the pump speed, v the piston speed, Pl as the pressure difference in the piston, and y the piston position signal.

For the closed-loop position control of the piston (see Figure 1a), we use a coarse proportional controller which helps the learning examples cover a significant part of system's working domain. The controller also attenuates the system dynamics dependency on secondary variables. This dependency is mainly visible when the learning examples are obtained from an open-loop system.

We built two learning sets containing examples of the electro-mechanical system behaviour: one set obtained with load applied to the piston (Figure 2a) and the other one obtained without load (Figure 2b). The learning sets, composed by the signals y, y_{ref}, ω, v and Pl, are acquired using a sinusoidal reference for the piston position with different amplitudes and frequencies. Another two sets of experimental values are used for testing the learning capability of the extracted fuzzy model. These sets considers the presence or not of any applied load to the piston, too.

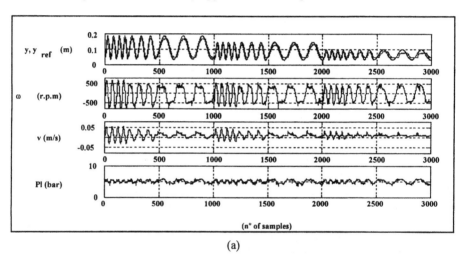

(a)

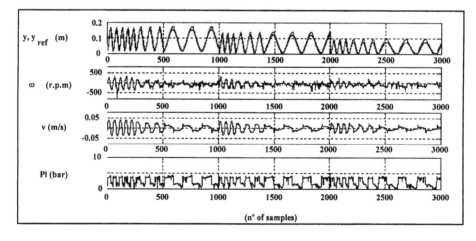

(b)

Fig. 2. Learning sets obtained without load applied to the piston (a) and with some load (b).

3 The Learning Process

The investigation of the proposed approaches uses a simple learning algorithm that utilises fuzzy logic to extract the system rules. In [3], the learning method is explained in details and an example illustrates the steps of the algorithm. In this section we present the main characteristics of the method and some results are presented applying the algorithm in extracting the fuzzy model the electro-mechanical system.

The learning process uses the traditional t-norm max to select the degree to which two fuzzy sets match. The method considers the most usual and simple structure where each variable is equally partitioned by membership functions of symmetric triangular shape. The fuzzy model uses rules as (1) where: $R^{(l)}$ is the lth rule, x_j denotes the variables expressing the system's dynamic condition, and y is the system output variable. The symbols $A_j^{(l)}$ represent the membership functions and $\omega^{(l)}$ are the rule consequent part, being a numerical value (*fuzzy singleton $B^{(l)}$*) considered as a pré-defuzzified output.

$$R^{(l)}: \text{if } x_1 \text{ is } A_1^{(l)} \text{ and } x_2 \text{ is } A_2^{(l)} \ldots \text{ and } x_m \text{ is } A_m^{(l)} \text{ then } y \text{ is } \omega^{(l)} \qquad (1)$$

The firing degree of each rule is measured using the expression (2). The variables denoted by $x_1', x_2' \ldots, x_m'$ represent the numerical values of $x_1, x_2 \ldots, x_m$, and the terms $\mu_{A_1^{(l)}}(.), \ldots, \mu_{A_m^{(l)}}(.)$ are the membership functions attributed to the fuzzy sets $A_1^{(l)}, \ldots, A_m^{(l)}$. The membership value of each variable, defined for the respective membership function composing the antecedent part of the rule (l), is combined with the other values by the *product operator* $(*)$.

$$\mu_{A_1^{(l)} \times \ldots \times A_m^{(l)}}(x_1, \ldots, x_m) = \mu_{A_1^{(l)}}(x_1')* \ldots *\mu_{A_m^{(l)}}(x_m') \qquad (2)$$

Figure 3 shows an example for a system with two input variables (x_1, x_2) and one output y. The figure illustrates the working domain partition by the fuzzy sets with each variable being quantized by a number of triangular and symmetric membership functions. The working domain is partitioned in subspaces where each one constitutes a rule describing a local relation. Each rule is a fuzzy implication statement with a degree calculated by expression (3), with the operation "$*$" being the inference product. As the consequent part is a fuzzy singleton set $B^{(l)}$, the value of $\mu_{B^{(l)}}(.)$ is 1.0 and the final degree is defined only by the multiplication of the antecedent membership values.

$$\mu_{A_1^{(l)} \times \ldots \times A_m^{(l)} \to B^{(l)}}(x_1', \ldots, x_m', y') = \mu_{A_1^{(l)}}(x_1')* \ldots *\mu_{A_m^{(l)}}(x_m')* \mu_{B^{(l)}}(y')$$
$$= \mu_{A_1^{(l)}}(x_1')* \ldots *\mu_{A_m^{(l)}}(x_m') \qquad (3)$$

We proceed as follows to derive, from learning data, the function that models the system. For each input x_i and output y_i, we calculate their different grades in the attributed membership functions. A set of n membership grades is attributed to each variable, corresponding to the number of membership functions or the number of linguistic terms of each variable. This grade set results in a vector with n possibilities. Then, for all variables, we choose a final degree that corresponds to the highest one in all possibilities values. The final membership functions, corresponding to the highest degrees, form a rule. As the learning data comes, we choose that rule with highest implication degree to calculate the pre-defuzzified output value $\omega^{(l)}$.

The inference method uses the centroid-defuzzification formula. It combines all rule contributions in a weighted form (equation 4) to compute the model output Y. The variable $\omega^{(l)}$ is the rule output, $\mu(R^{(l)})$ is the rule firing degree and c corresponds to the number of extracted rules.

$$Y = \frac{\sum_{l=1}^{c} \mu(R^{(l)}).\omega^{(l)}}{\sum_{l=1}^{c} \mu(R^{(l)})} \tag{4}$$

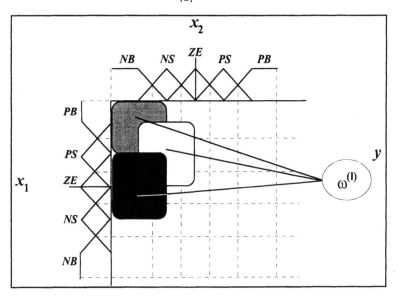

Fig. 3. Example of a working domain partition by fuzzy sets for a system with two input variables (x_1, x_2) and one variable $\omega^{(l)}$.

The results presented below show the generalisation capability of the extracted model. Considering our theoretical knowledge about the process, we establish the relation (5) to represent the system. The relation expresses the piston position (y) as a function of the reference signal (y_{ref}), the piston speed (v) and the pump speed (ω).

$$y = f(y_{ref}, v, \omega) \tag{5}$$

Using the first training set, we apply the learning process to extract the fuzzy model. Figure 4 shows the error evolution between the fuzzy model prediction and the measured piston position. In spite of the good approximation obtained for the extracted relation (5), Figure 4 shows high oscillations in the error signal from the learning process. The errors may originate from the incomplete information supplied in the learning process, from noise interference, and from the learning data not cover a significant part of the system working domain.

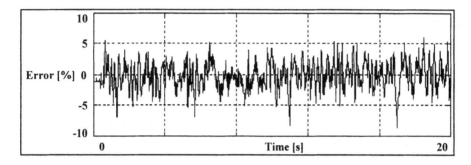

Fig. 4. Position error signal evolution of fuzzy model.

It is essential to know what variables are the most influent in the system dynamics. Some technique that measures the correlation between variables, as the PCA, helps to indicate the variables more important to describe the system dynamics.

For the learning algorithm, we can consider generalisation a local effect. Therefore, the learning data must cover a significant part of the system working domain. If there are regions that not receive experimental data and, which extracted rules have a zero conclusion, they are also considered in the learning process.

The pre-processing method using PCA permits us to enlarge the disposable information for fuzzy learning process resulting in more robust fuzzy models. Figure 5 presents the position error evolution obtained after the application of PCA.

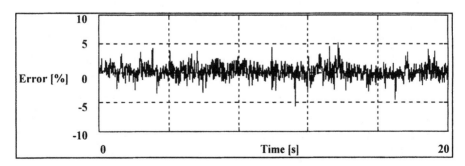

Fig. 5. Error signal evolution after the pre-processing method PCA.

4 Structure Identification I: Selecting the Input Variables

The first step in structure identification is identify the input variables that affect the system output. If the number of input variables is high, it is convenient to reduce our attention to a smaller group and so work with acceptable rule-bases.

Some methods, as for example the regularity criterion [6], have been used to indicate the best input variables. In this section, we use the statistical method of Principal Component Analysis to indicate the variable set dominating the system dynamics. This technique permits to enlarge the available information used to extract the fuzzy model increasing its performance.

4.1 Principal Component Analysis

Principal Component Analysis is a technique that seeks to describe the multivariate structure of the data [8]. In order to examine the relationships among a set of p variables, PCA transforms the original set of variables to a new set of uncorrelated variables called *principal components*.

The start point for PCA is the sample covariance matrix S calculated for a p-variable problem by (7). The term s_i^2 (8) is the variance of ith variable x_i, and s_{ij} (9) denote the covariance between ith and jth variables with $\bar{x}_i$ and $\bar{y}_i$ being the mean value of x and y. In equations (8) and (9), n is the number of data values of the learning set.

$$S = \begin{bmatrix} s_{11}^2 & s_{12} & \cdots & s_{1p} \\ s_{12} & s_{22}^2 & \cdots & s_{2p} \\ \vdots & \vdots & & \vdots \\ s_{1p} & s_{2p} & \cdots & s_{pp}^2 \end{bmatrix} \tag{7}$$

$$s_i^2 = \frac{1}{n-1}\sum_{i=1}^{n}(x - \bar{x}_i)^2 \tag{8}$$

$$s_{ij} = \frac{1}{n-1}\sum_{i,j=1}^{n}(x - \bar{x}_i)(y - \bar{y}_j) \tag{9}$$

The principal axis transformation is illustrated in Figure 6. The transformation converts the p variables $\vec{x} = (x_1, x_2, ..., x_p)$ into a p new uncorrelated variables $\vec{z} = (z_1, z_2, ..., z_p)$ called principal components. These new variables are linear combinations of the original ones and are derived in decreasing order of importance so that, for example, the first component accounts for as much as possible for the data variation.

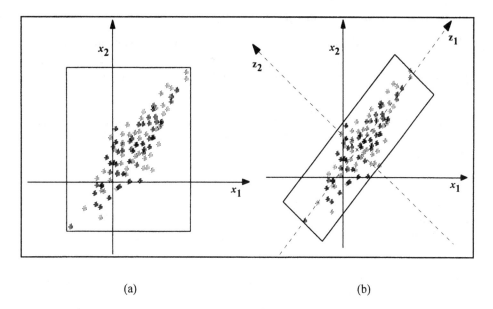

(a)   (b)

Fig. 6. (a) Data and the domain defined by the original coordinates (x_1, x_2). (b) Transformation of the coordinates made by PCA. The new domain is defined by the principal components (z_1, z_2).

The eigenvectors $\vec{u}_i$ of the covariance matrix S describe the new system coordinates. They establish a matrix U of direction cosines which is used in the transformation of the coordinates $\vec{x}$ to $\vec{z}$, as described in (10).

$$\vec{z} = U^T \vec{x} \tag{10}$$

The variables z_i have zero mean and a variance value λ_i. The variance corresponds to the ith eigenvalue computed from matrix S. The linear combination between the ith eigenvector $(\vec{u}_i)$ and the original variables $(\vec{x})$ calculate every ith principal component, as shown in expression (11).

$$\vec{z}_i = \vec{u}_i^T \vec{x} \tag{11}$$

If the first principal components account for most of the variation in the original data, then it is argued that the effective dimensionality of the problem is less than p. In other words, if some of the original variables are highly correlated, they are effectively "saying the same thing" and one of the variables is redundant within the set. The key to this detection lies in the vector associated with the highest eigenvalue.

4.2 Denoting the Most Significative System Variables using PCA

The analysis of a learning set using the technique of PCA permits investigate the relative contribution of each candidate variable to characterise the systems dynamic. The analysis uses the information contained in the computed eigenvectors $\vec{u}_i$ and respective eigenvalues λ_j.

Following, the pre-processing is applied to each learning set. We introduce some results concerning the selection of the system variables and the benefits of using the principal components in a fuzzy modeling process.

Learning Set Without Load Applied to the Piston.

We calculate the covariance matrix from the learning set. Table I shows the computed eigenvalues and respective eigenvectors. We put in evidence the main eigenvector and respective eigenvalue.

The eigenvalues measure the variance of each principal component z_i. The highest eigenvalue ($\lambda_1 = 8.3648$) is attributed to z_1 that accounts for almost 88% of total data variance. Each variable is weighted by the respective eigenvector component. In equation (12), we show the computation of the first component z_1 as a linear combination of the original variables.

$$z_1 = (+0.7694)y_{ref} + (0.0460)v + (0.0170)Pl + (0.0809)\omega + (0.6317)y \qquad (12)$$

Table I					
Eigenvalues	8.3648	0.9057	0.1023	0.0034	0.0010
Eigenvector	u_1	u_2	u_3	u_4	u_5
y_{ref}	0.7694	-0.3174	0.2726	-0.4769	-0.0739
v	0.0460	-0.5978	-0.7963	0.0045	0.0806
Pl	0.0170	-0.0746	0.1567	0.0140	0.9846
ω	0.0809	-0.5345	0.3984	0.7319	-0.1157
y	0.6317	0.5006	-0.3292	0.4865	0.0725

In Table II, we show the eigenvector components for the highest eigenvalue. We see that each variable has a weighting eigenvector component. The set of system variables is divided into a group of antecedent variables (y_{ref}, ω, v, Pl) and the system output variable, the piston position y. We obtain a global order of importance into the variable set taking the absolute value of each eigenvector component.

Table II shows the ordering of the four antecedent variables. The reference signal y_{ref} is the most significative variable, ω and v are the followings, and Pl the last significative variable. This happens because the pressure signal is closely related with the presence, or not, of the load. PCA indicates that the pressure signal has no useful information to the learning. The variable is not significative, in this case, to describe the system dynamics.

Table II				
Variable's global ordering				
y_{ref}	y	ω	v	Pl
0.7694	0.6317	0.0809	0.0460	0.0170
Ordering of the antecedent variables				
y_{ref}		ω	v	Pl

Learning Set With a Random Variable

For another test, we insert into the five variables a random variable denoted RAND. Table III shows the eigenvectors and respective eigenvalues from computed covariance matrix.

The random variable is not correlated with the other system variables. This fact appears in the principal eigenvector through the component related with the RAND variable. The component is the smallest among all components with an absolute value of 0.0033, registering the ambiguity of the RAND variable in the variable set.

We can notice too the ambiguity by the smallest eigenvalue and observe that the highest component of its eigenvector, with value of 0.9984, is attributed to the RAND variable. The random variable maintains the global ordering since it is not correlated with any other variable in the set.

Table III						
Eigeinvalue	8.3649	0.9057	0.1837	0.1023	0.0034	0.0010
Eigeinvector	u_1	u_2	u_3	u_4	u_5	u_6
y_{ref}	0.7694	0.3174	0.2726	0.4769	0.0739	0.0039
v	0.0460	0.5978	-0.7962	-0.0045	-0.0806	-0.0074
Pl	0.0170	0.0746	0.1567	-0.0141	-0.9846	0.0017
ω	0.0809	0.5345	0.3984	-0.7319	0.1170	0.0012
RAND	-0.0033	0.0029	-0.0074	-0.0005	0.0007	0.9984
y	0.6317	-0.5006	-0.3293	-0.4865	-0.0725	0.0009

Learning Set with Load Applied to the Piston

The eigenvectors and eigenvalues obtained from this learning set are shown in Table IV.

Table IV					
Eigenvalue	10.4096	0.7164	0.0692	0.0105	0.0047
Eigenvector	u_1	u_2	u_3	u_4	u_5
y_{ref}	0.6910	0.2079	0.5512	0.2965	0.2959
v	0.0200	0.9219	-0.1908	-0.0024	-0.3366
Pl	0.0356	0.1804	-0.5553	0.0135	0.8109
ω	-0.0122	0.1447	0.3344	-0.9066	0.2124
y	0.7216	-0.2311	-0.4895	-0.2999	-0.3105

Table V shows the new ordering established by the most significative eigenvector u_1. Comparing Table V with Table II, it is noticed that, if we look for the highest eigenvalue $\lambda_1 = (10.4096)$, the global ordering has changed. The piston position (y) continues to be the more important variable. The pressure (Pl) passes to be more significant signal compared with motor speed (ω) and piston speed (v) in the antecedent set. This indicates the high correlation between the pressure signal and the load applied to the piston.

Table V				
Variable's global ordering				
y	y_{ref}	Pl	v	ω
0.7216	0.6910	0.0356	0.0200	0.0122
Ordering of the antecedent variables				
y_{ref}	Pl		v	ω

4.3 Fuzzy Modeling Using the Principal Components

The set of eigenvectors and eigenvalues computed from the covariance matrix S permits to obtain a set of uncorrelated variables denoted by principal components. These new coordinates, obtained after an orthogonal rotation made by the matrix U, introduces a new working domain where all variables have maximal variance. Figure 6 showed how the transformation of the coordinates can be advantageous to a fuzzy modeling process. The learning examples fill better the new domain and decrease the possibility of appearing empty rules in the fuzzy model. The fuzzy model complexity decreases because the number of identified system rules is reduced. The information about system behaviour is extended as the new coordinates have high variance.

Therefore, the performance of the learning process increases because uses more information than before.

The change of coordinates can be performed in the antecedent and consequent variables or only in the antecedent ones. At follow, we investigate these two cases in fuzzy modeling of the electro-hydraulic system.

Change of Coordinates Using The Antecedent Variables

The function (13) approximates the electro-mechanical system model. The piston position is considered as a function of the reference signal, the pump speed, the piston speed and the pressure difference in the piston.

$$y = f(y_{ref}, \omega, v, Pl) \qquad (13)$$

From the previous section, the pressure signal Pl can be neglected for a no-load system, reducing the function to three variables (14).

$$y = f(y_{ref}, \omega, v) \qquad (14)$$

Using the no-load learning set, we make a transformation in the domain established by the three antecedent variables. We compute the eigenvectors and eigenvalues of the covariance matrix and construct the principal components. Table VI presents their values and point out the highest eigenvalue.

Table VI			
Eigenvalue	5.1974	0.5259	0.0662
Eigenvector	u_1	u_2	u_3
y_{ref}	0.9844	0.0555	-0.1668
v	0.1003	0.6020	0.7922
ω	0.1443	-0.7965	0.5871

The linear combination of the variables with the computed eigenvectors u_1, u_2, u_3 converts the antecedent variables into the principal components. The equation (15) presents the principal components with respective eigenvalues.

Figure 7 shows the axis coordinates before and after the transformation made by PCA. The axe of piston speed (v) rotates to z_2, the axe of motor speed (ω) rotates to z_3 and the piston position reference y_{ref} rotates a small angle in direction of z_1.

$$\begin{cases} z_1 = 0.9844 * y_{ref} + 0.1003 * v + 0.1443 * \omega & \lambda_1 = 5.1974 \\ z_2 = 0.0555 * y_{ref} + 0.6020 * v - 0.7965 * \omega & \lambda_2 = 0.5259 \\ z_3 = -0.1668 * y_{ref} + 0.7922 * v + 0.5871 * \omega & \lambda_3 = 0.0662 \end{cases} \qquad (15)$$

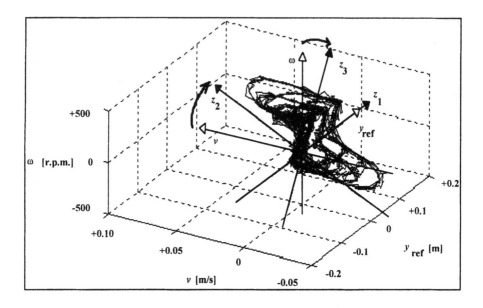

Fig. 7. Change of domain coordinates $(y_{ref}, \omega, v) \rightarrow (z_1, z_2, z_3)$

To better visualise the new domain, the tri-dimensional data space of Figure 7 is projected into three bi-dimensional planes. Figures 8 and 9 present two plane projections before and after coordinates change. In both figures, the new domain defined by z coordinates is better filled, making the before concentrated information to be spanned.

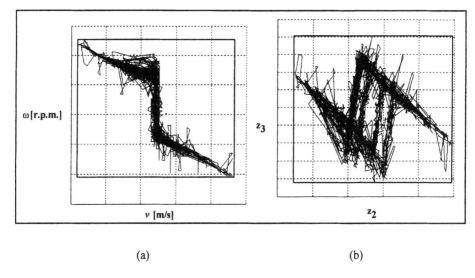

(a)   (b)

Fig. 8. First plane projection for change of coordinates $(v, \omega) \rightarrow (z_2, z_3)$. (a) System coordinates before PCA presenting condensed information in the working domain. (b) After transformation of the coordinates, the new domain presents more information than before.

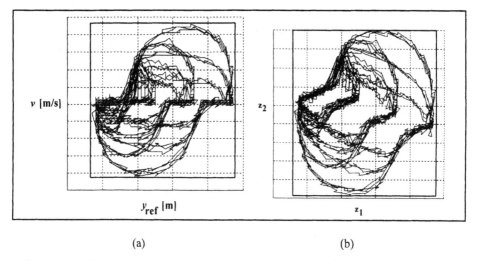

(a)   (b)

Fig. 9. Second plane projection for change of coordinates $(y_{ref}, v) \rightarrow (z_1, z_2)$. (a) System coordinates before PCA presenting in the central region an accumulation of data. (b) After PCA the central data is spanned.

Figure 10 shows the third data projection. The domain, specified by the original system coordinates (y_{ref}, ω), is very sparse (see Figure 10a), making high probable the appearing of empty rules in the extracted fuzzy model. After the change of coordinates, the learning data is displaced in the manner to improve the covering of the domain (Figure 10b).

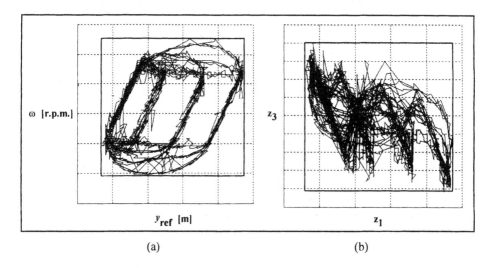

(a)   (b)

Fig. 10. Third plane projection for change of coordinates $(y_{ref}, \omega) \rightarrow (z_1, z_3)$. (a) System coordinates presenting a sparse domain. (b) After transformation the data is reallocated decreasing the sparse domain.

The uncorrelated variables obtained by PCA create an improved learning set and increases the available information to the learning process. So, the principal components can be the new antecedent variables replacing the anterior ones (16) and forming an alternative fuzzy model.

$$(y_{ref}, \omega, v) \rightarrow y \quad \langle \text{original model} \rangle$$
$$\Downarrow \tag{16}$$
$$(z_1, z_2, z_3) \rightarrow y \quad \langle \text{modified model} \rangle$$

4.3 Fuzzy Modeling Using the Principal Components as the New Antecedent Variables

The new coordinates are utilised by the learning algorithm explained in section 3. The number of membership functions is computed by the "trial-and-error" process to minimise the quadratic error mean E denoted in (17). The term y_i stands for the system output and Y_i the respective fuzzy model output.

$$E = \frac{1}{n} \sum_{i=1}^{n} (y_i - Y_i)^2 \tag{17}$$

Modeling Using the Original Coordinates.

The number of membership functions obtained for each variable is indicated in (18). The extracted fuzzy model presents a quadratic error of $E = 0.2539$. Figure 11 presents the error signal evolution between the system and the model output.

$$
\begin{array}{llll}
y_{ref} & \rightarrow & (11) & \text{membership functions} \\
v & \rightarrow & (11) & " \\
\omega & \rightarrow & (11) & " \\
y & \rightarrow & (13) & "
\end{array}
\tag{18}
$$

Modeling Using the Principal Components as the Antecedent Variables.

The new coordinates define a new learning set for which another fuzzy partition has to be computed. In (19), we show the partitions attributed to each new coordinate. Testing the extracted fuzzy model, the quadratic error E decreases by 14% compared with the anterior value. The total of fuzzy partitions decreases about 66% so reducing significantly the number of rules need to represent the process.

$$
\begin{array}{llll}
z_1 & \rightarrow & (13) & \text{membership functions} \\
z_2 & \rightarrow & (7) & " \\
z_3 & \rightarrow & (5) & " \\
y & \rightarrow & (13) & "
\end{array}
\tag{19}
$$

Modeling Using the Principal Components in Antecedent and Consequent Variables.

Principal Component Analysis is applied to the variable set composed both by antecedent variables (y_{ref}, v, ω) and the consequent one (y). Table VII shows the computed eigenvectors and respective eigenvalues.

Table VII				
Eigenvalue	8.3624	0.9007	0.0998	0.0034
Eigenvector	u_1	u_2	u_3	u_4
y_{ref}	0.7695	-0.3190	0.4758	-0.2824
v	0.0459	-0.6007	-0.0033	0.7982
ω	0.0808	-0.5355	-0.7335	-0.4107
y	0.6319	0.5006	-0.4854	0.3384

The equation set (20) shows the principal components and respective eigenvalues.

$$
\begin{aligned}
z_1 &= 0.7695 * y_{ref} & +0.0459 * v & +0.0808 * \omega & +0.6319 * y & \quad \lambda_1 = 8.3624 \\
z_2 &= -0.3190 * y_{ref} & -0.6007 * v & -0.5355 * \omega & +0.5006 * y & \quad \lambda_2 = 0.9007 \\
z_4 &= 0.4758 * y_{ref} & -0.0033 * v & -0.7335 * \omega & -0.4854 * y & \quad \lambda_3 = 0.0998 \\
z_3 &= -0.2824 * y_{ref} & +0.7982 * v & -0.4107 * \omega & +0.3384 * y & \quad \lambda_3 = 0.0034
\end{aligned} \tag{20}
$$

The fuzzy rules are now composed by the antecedent variables z_1, z_2, z_3 and, as the new consequent one, the component z_4 instead of variable y. Using the new relation $(z_1, z_2, z_3) \rightarrow z_4$ into the learning process, the extracted model obtains a quadratic error of $E = 0.0933$. The partition number established for this model is shown in (21).

$$
\begin{aligned}
z_1 &\rightarrow (7) \quad \textit{membership functions} \\
z_2 &\rightarrow (3) \quad \quad " \\
z_3 &\rightarrow (5) \quad \quad " \\
z_4 &\rightarrow (9) \quad \quad "
\end{aligned} \tag{21}
$$

We compare the results with those obtained using the original system variables without change of coordinates. The number of fuzzy partitions decreases about 94% and the quadratic error E decreases 63%.

If we compare the results with those using change of coordinates only in the antecedent variables, the partition decreases 84% and the quadratic error E decreases 56%.

Figure 11 shows the error signal when we use the learning process without pre-processing by PCA. Figure 10 shows the error signal obtained with the new fuzzy model.

The results are associated with four points:

- the change of the coordinates makes the working domain less sparse allowing us to work with smaller fuzzy partitions for each variable;

- the new domain reduces the inference errors caused by possible rules with a null conclusion;

- more information is available from the new learning set with a consequent learning improvement, decreasing significantly the quadratic error of the fuzzy model;

- the new variables are desacoupled originating more robust rules;

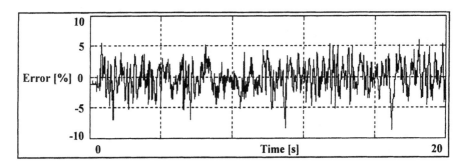

Fig. 11. Error signal without change of coordinates. (11-11-11-13, E=0.2539)

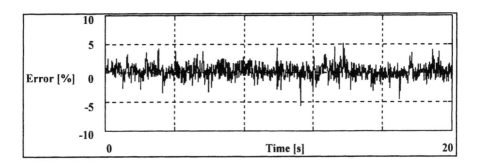

Fig. 12. Error signal evolution after the change of coordinates. (7-3-5-9, E=0.0933)

Summary of the Steps of the Fuzzy Modeling Algorithm Using the Principal Components

The final algorithm is resumed by two steps.

Step 1.: This step concerns the learning process to extract the linguistic model of the electro-hydraulic system. The PCA method transforms the system variables (y_{ref}, v,

ω, y) in the new variables (z_1, z_2, z_3, z_4). With the new learning set, the learning algorithm extracts rules like (22).

$$\text{if } (z_1, z_2, z_3) \text{ then } (z_4) \tag{22}$$

Step 2.: After the learning process, the inferred rules must be tested in their generalisation capability. We use a test set with examples not presented in the learning phase.

For each systems condition (y_{ref}, v, ω), the new coordinates are calculated using the equations that define the principal components. Although, it must be noted that the variables (z_1, z_2, z_3) are dependent of the system output that we can estimate. To surpass this situation, the variables are initially calculated without the output signal, with $y = 0.0$. The output coordinate z_4 is inferred by the learned fuzzy rules.

Using the equation (23), the inferred value z_4 and the signals (y_{ref}, v, ω) are used to estimate the system output y. With that operation, it will have an error which deviates y from the correct value because we do not use it for computing (z_1, z_2, z_3). Then, the output value inferred before can now be used in the recalculation of (z_1, z_2, z_3) giving us a more correct output.

$$z_4 = 0.4758 * y_{ref} - 0.0033 * v - 0.7335 * \omega - 0.4854 * y \tag{23}$$

5 Structure Identification II: Fuzzy Partition of Input Space

Part I of structure identification is concerned in identify the variables that better characterise system dynamics. This subset is separated into a group of input variables that composes the antecedent rule part, and an output variable being the consequent variable. As the consequent variable in the fuzzy rule-base is of singleton type, part II of structure identification consists in estimate the fuzzy partition of input space. This partition involves the number of membership functions attributed to each variable, their position in the universe of discourse and their fuzziness degree, expressed by the fuzzy set's width.

We present in this section the application of the *autonomous mountain-clustering method* [2] in identify the premise structure. The algorithm analyses the learning data computing a set of clusters which describes the data structure. Figure 13 shows an example for a bi-dimensional input space. The clusters are projected onto each axe, so obtaining a membership function localised at the respective cluster coordinate. The width of the membership functions is taken as the distance between the adjacent clusters and, for simplicity, we use the triangular function as the fuzzy set used. Originating non-symmetrical membership functions, each side of the membership function expresses the compactness or dispersion of the data around the considered cluster in each axes direction. This characterises the data distribution in the input space.

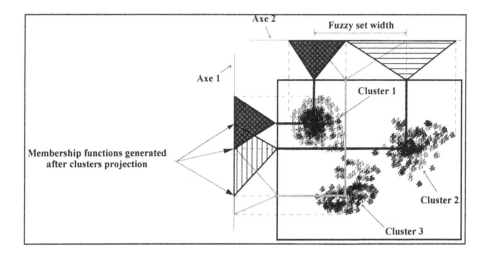

Fig. 13. Figure that illustrates the identification of data set structure. The projection of computed clusters generates membership functions in each universe of discourse.

5.1 Autonomous Mountain-Clustering Algorithm

The *autonomous mountain-clustering algorithm* is divided into four steps:

- In the first step, the input space is discretized by a grid to generate candidate cluster centers;

- The second step attributes to each candidate cluster a density value based on a parameter set inferred automatically from data. These collection of grades compose a potential function that, plotted for a bi-dimensional case, is analogous to a mountain;

- The third step uses an iterative process for the mountain function reduction ("peeling" process) and computes the best cluster centers. As the centers are fixed by the grid, its precision is highly dependent on the initial established grid granularity;

- The final step of the cluster algorithm uses an iterative reallocation procedure to make a fine correction on each cluster.

In the following, we explain in detail the autonomous algorithm. For a better understanding of each step, the algorithm is applied to a bi-dimensional data set that is displayed in Figure 14.

We consider a data set generated by an electrical drive system implemented in the laboratory. The data is composed by the speed of an induction motor and its corresponding stator current with both signals scaled to a tension value. The signals, contaminated by noise, have a simplistic relation as seen by Figure 14. Although, this simplistic relation permits us to easily show in tri-dimensional space the mountain function, the "peeling" process, and the cluster's reallocation process.

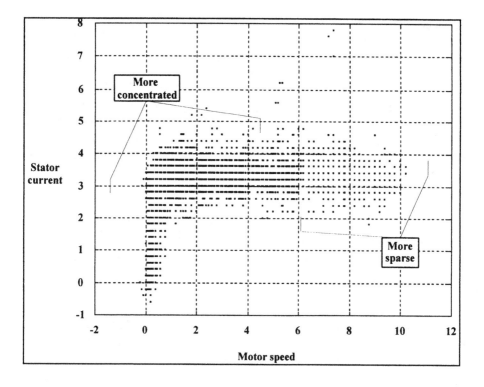

Fig. 14. Figure displaying the data set used to illustrate the autonomous clustering process.

First Step.

Input-Space Discretization: Let's consider an *n*-dimensional data space with N_p data points and the corresponding *n*-dimensional cube that holds all data. In the example, the data space has $n = 2$ and it is contained in a square.

The first step divides each variable range by a number of partitions (N_q), so that the data range is broken into $(N_q)^n$ squares. The partition number is supposed to be equal for all variables.

Because the data values are considered to be independent or equally important, we can consider the data being the result of a binomial distribution. As the number of collected data values is much smaller than the total number of disposable data values, they can be represented by a Poisson distribution [7]. The advantage of considering this distribution is that the number of spurious points in the data set is equal to the square root of the total number of data points. This means that if we want to have simplicity of the system's representation without lost much information, the average number of points per cluster must be around the number of spurious points calculated by $\sqrt[2]{N_p}$.

The number of candidate clusters considering *n* variables with the same partition number is $(N_q)^n$. Therefore, the number of partitions (N_q) must be settled around the value of $2\sqrt[n]{N_p}$ as

$$(N_q)^n = \sqrt[2]{N_p}.$$

Following with the example, the data measured has $N_p = 10052$ values giving us a "reference value" for the partition number of approximately 10. The anterior assumptions consider the data equally distributed by the domain. In practice, the data is not uniform. So, it is perceived that a number of 8 partitions for each variable is enough to give us a good cluster representation of the data and losing few information.

Second Step.

Attributing a Density Value to Each Cluster Candidate: A grid, formed by the $N_q \times N_q$ partitions, discretises the input space. The vertices produced by the grid lines intersection makes the set of candidate clusters. Each cluster candidate (i) is denoted as N_i, with $1 \le i \le (N_q)^n$.

A density value (M) is attributed to each cluster candidate N_i. The density value is computed adding to each candidate N_i a certain quantity coming from each data value. This amount represents the contribution grade of that value to the vertice election as a good cluster.

The weight of each data value to N_i decreases as the distance between them increases. To represent this relation, an exponential function with a decaying parameter α is employed. Each density value is calculated by equation (24). The notation $M^1(N_i)$ denotes the density value M obtained from the N_p data points for the cluster candidate N_i, and the indice 1 indicates that it is related with the construction of the first "mountain function". The expression $e^{-\alpha|P_k - N_i|}$ measures a certain "membership grade" of the data value P_k in the cluster candidate N_i, weighted by the Euclidean distance $|P_k - N_i|$ and a decaying parameter α [1].

$$M^1(N_i) = \sum_{k=1}^{N_p} e^{-\alpha|P_k - N_i|} \tag{24}$$

Setting the Parameter α: The parameter α can be fixed by the operator through a "trial-and-error" method, as initially proposed by Yager and Filev [1]. This process takes much time and the final clusters have to be bi-dimensional or three-dimensional displayed together with the data values. To surpass that, we proposed a solution in [2] which automatically computes the parameter α from the data set.

Let's consider, for simplicity, the case of a bi-dimensional data set (x,y). The generalisation of the results to higher dimensions is straightforward. The variable x is considered between x_0 and $x_0 + L_1$, and y between y_0 and $y_0 + L_2$. The grid lengths in x and y directions, respectively d_x and d_y, are the division of their ranges by the partition number N_q (25).

$$d_x = \frac{L_1}{N_q}, \ d_y = \frac{L_2}{N_q} \tag{25}$$

For small grid lengths, the exponential term in the density equation (24) has to decay faster since data is supposed to be more concentrated. In the other side, if the data is sparse, the grid lengths will be larger and the exponential term has to decay slowly. Then, it is necessary analyse the data distribution in the two directions x and y.

As the parameter α is responsible by the magnitude of the exponential decay, it can not have the same value for both directions. The scalar parameter becomes a vector $\vec{\alpha}$ with two components, α_x and α_y, computed as the inverse of grid partitions d_x and d_y, as stated in equation (26).

$$\vec{\alpha} = (\alpha_x, \alpha_y) = \left(\frac{1}{d_x}, \frac{1}{d_y} \right) \tag{26}$$

Observing Figure 15, each data P_k contributes with two different amounts for the density value M of the candidate cluster N_i. One amount (29) comes from the relative position of P_k to N_i, considering as reference the x direction The other amount (30) considers as reference the y direction.

Each amount is weighted by the respective $\vec{\alpha}$ component related with x or y direction, and is composed by the projection of the distance r_{ki} (27) between the candidate cluster $N_i = (N_{i_x}, N_{i_y})$ and the data value $P_k = (x_k, y_k)$.

The exponential term in equation (24) is rewritten in (28) taking into account the angle $(_i\theta_k)$ between P_k and N_i. The equation (28) uses the square of cosine and sine so that when we assume $\alpha_x = \alpha_y = \alpha$ the original equation (24) is recovered.

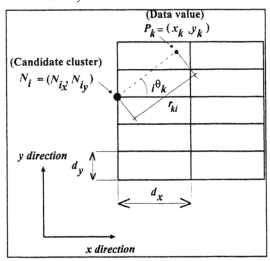

Fig. 15. Grid length's and the angle between cluster N_i and a data value (x_k, y_k).

$$r_{ki} = |x_k - N_i| = \sqrt{(x_k - N_{i_x})^2 + (y_k - N_{i_y})^2} \tag{27}$$

$$e^{-(\alpha_x \cos^2(_i\theta_k) + \alpha_y \sin^2(_i\theta_k))r_{ki}} \tag{28}$$

The terms $\cos(_i\theta_k)$ and $\sin(_i\theta_k)$ in (28) are calculated using expressions (29) and (30).

$$\cos(_i\theta_k) = \frac{(x_k - N_{i_x})}{r_{ki}} \tag{29}$$

$$\sin(_i\theta_k) = \frac{(y_k - N_{i_y})}{r_{ki}} \tag{30}$$

Substituting the equations (29-30) in (27), the new ponderation term results in equation (31). The density value is computed by the sum over all N_p data values as shown in equation (32).

$$e^{-\left(\frac{\alpha_x(x_k - N_{i_x})^2 + \alpha_y(y_k - N_{i_y})^2}{\sqrt{(x_k - N_{i_x})^2 + (y_k - N_{i_y})^2}}\right)} \tag{31}$$

$$M^1(N_i) = \sum_{k=1}^{N_p} e^{-\left(\frac{\alpha_x(x_k - N_{i_x})^2 + \alpha_y(y_k - N_{i_y})^2}{\sqrt{(x_k - N_{i_x})^2 + (y_k - N_{i_y})^2}}\right)} \tag{32}$$

After the calculation of all density values, we can construct the "mountain function". Figure 16 shows the "mountain function" obtained for the data set illustrated in Figure 14. The domain is defined by the set of all candidate clusters and the image is defined by the real positive density values M. The best cluster is associated with the mountain peak, which has the highest density value. The other values of the mountain function measure the potentiality of each grid point to be a cluster and provides an *image* of the data distribution.

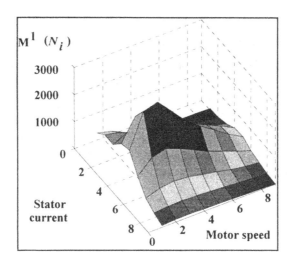

Fig. 16. Mountain function

Third Step.

Using the Mountain Function to Compute the Best Clusters: The mountain peak is the point with highest density and is the principal cluster. To localise the second best cluster, it is necessary to eliminate the effect of the principal cluster from the other cluster candidates. For grid vertices closer to the mountain peak, their density value must be more attenuated than for those further away. This process generates another mountain function, from which the peak is elected the second best cluster. Again, the "peeling" process is repeated to find the next cluster and so on.

We denote by N_1^* the peak of the first mountain with highest density value M_1^*. The density values of the second mountain function are denoted by M^2 and their values are computed by equation (33). The equation (33) eliminates the effect of the cluster N_1^* over the other cluster candidates, localising the second better cluster.

The parameter β of equation (33) is called the mountain elimination parameter. Its function is similar to the parameter α but now associated with the relative localisation of the clusters [1]. In the "peeling" process, the parameter β is dependent, at each iteration k, on the distance between the main cluster and the candidates.

$$M^2(N_i) = M^1(N_i) - M_1^* e^{-\beta|N_1^* - N_i|} \qquad (33)$$

Setting the Parameter β: At each iteration k, the parameter β depends on the principal cluster's position relative to the considered grid vertice (being a cluster too but with a smaller density value). This consideration introduces a parameter vector

$\vec{\beta}(k)$ with two components as the parameter $\vec{\alpha}$. However, the parameter β becomes different at each iteration because the reference cluster $N^{*}(k)$ changes too.

The cluster candidate at iteration k is composed by two coordinates $N_i(k) = (N_{i(k)_x}, N_{i(k)_y})$, and the cluster taken as the reference at the same iteration by $N^{*}(k) = (N^{*}_{(k)_x}, N^{*}_{(k)_y})$. The equations (34-35) show the components of $\vec{\beta}(k)$. The differences $N^{*}_{(k)_x} - N_{i(k)_x}$ and $N^{*}_{(k)_y} - N_{i(k)_y}$ are, respectively, the x and y distances between the reference cluster and the candidate at iteration k. They have a similar role as d_x and d_y defined for parameter α.

$$\beta_x(k) = \frac{1}{(N^{*}_{(k)_x} - N_{i(k)_x})} \tag{34}$$

$$\beta_y(k) = \frac{1}{(N^{*}_{(k)_y} - N_{i(k)_y})} \tag{35}$$

"Peeling" the Mountain Function: The mountain function is reduced at iteration $k+1$ using the parameter vector $\vec{\beta}$. In the revised mountain reduction equation (36), $M^k(N_i)$ stands for the value of the mountain function on iteration k at the vertice N_i, M^{*}_k is the density value of the considered principal cluster at iteration k and $M^{k+1}(N_i)$ is the new mountain value after discounting the influence of other cluster candidates.

$$M^{k+1}(N_i) = M^k(N_i) - M^{*}_k e^{-\left(\frac{\beta_x(k)(N^{*}_{(k)_x} - N_{i_x})^2 + \beta_y(k)(N^{*}_{(k)_y} - N_{i_y})^2}{\left((N^{*}_{(k)_x} - N_{i_x})^2 + (N^{*}_{(k)_y} - N_{i_y})^2\right)^{\frac{1}{2}}} \right)} \tag{36}$$

At follow, we illustrate the "peeling" process. Figure 16 presents the first mountain function using 8 partitions for each variable. The mountain is reduced extracting its peak and is plotted in Figure 17. We see by the figure that some density values after the discount can be negative. These clusters are set to zero and eliminated as valid cluster candidates.

Figures 18 to 21 illustrate the evolution of the "peeling" process.

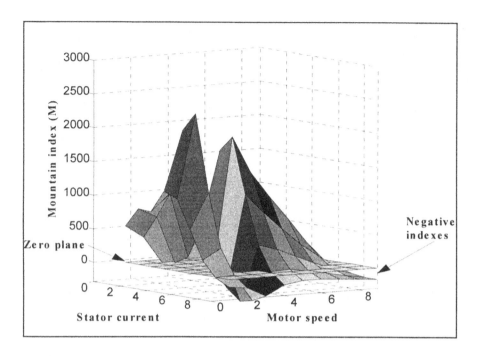

Fig. 17. Second mountain function generated after removing the first peak. It shows that some mountain indexes can be negative after the discount process

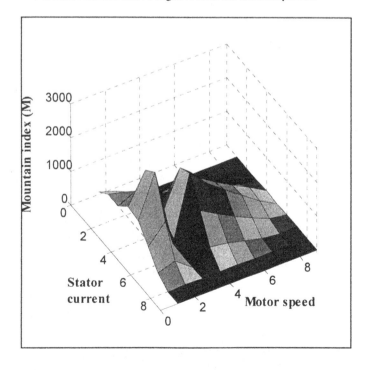

Fig. 18. First mountain reducion zeroing the negative indexes.

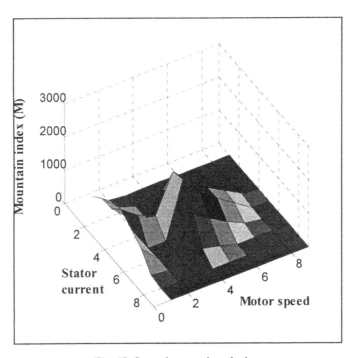

Fig. 19. Second mountain reducion.

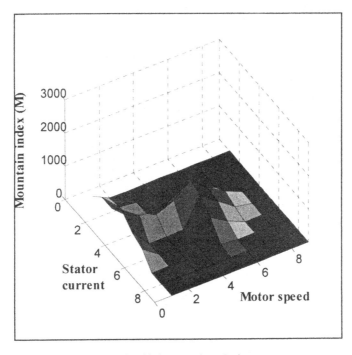

Fig. 20. Third mountain reducion.

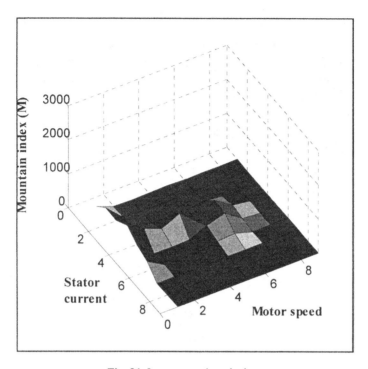

Fig. 21. Last mountain reducion.

Some Procedures to Stop the "Peeling" Process: The autonomous mountain-clustering algorithm considers two stopping procedures for the "peeling" process. It stops if the number of identified clusters surpasses the established value $2\sqrt[n]{N_p}$, or when there are no more valid candidates. The process reduction continues until all clusters have a mountain indice zero after the discount.

Figure 22 shows the clusters obtained after the "peeling" process.

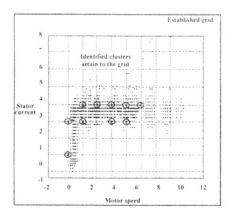

Fig. 22. Data set and the identified clusters, attained to the grid, using a number of 8 partitions for each variable.

Final Step

Associating a Window to Each Cluster: When the algorithm starts, all cluster candidates are fixed on the initial established grid vertices. Thus, even if we start with a great number of partitions, it is not completely true that the best clusters coincide with the established grid vertices. To correct that problem, we reallocate each cluster so that they get closer to the true clusters.

The reallocation process starts associating a window to each cluster. Figure 23 shows an example with 3 clusters. The clusters N_2 and N_3 are localised in the grid vertices with the complete associated windows. The cluster N_1 defines a special case where the cluster is localised in the border of the domain and the window is only defined partially.

The dimensions of the window are chosen as the grid partitions d_x and d_y, and the cluster is localised in the window center. The gravity-center of the window is calculated using the axis localised in the respective cluster as reference coordinates. All distances are between the cluster and the data into its associated window. The cluster is denoted N_i as before and each data denoted by m_{ki}. The symbol m_{ki} means the data contained into window i at iteration k.

To compute the gravity-center, we attribute to each data in the window a weight. This weight is an exponential measure of data contribution for the cluster localised at the gravity-center of the window. It is inversely proportional to the distance r_{ki} between the data k and the cluster i and uses as decaying factor the vector parameter $\vec{\alpha}$. The gravity-center is recalculated being its new coordinates given by the vector $\vec{Z}_{ki} = (z_1^{ki}, z_2^{ki})$ defined in (37).

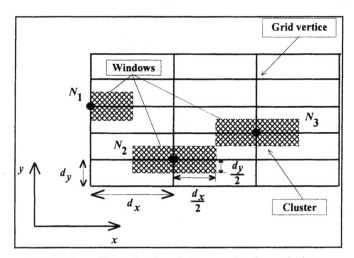

Fig. 23. Figure illustrating the window associated to each cluster.

$$z_1^{ki} = \frac{\sum_{t=1}^{m_{kj}} e^{-\alpha_x r_{ki}(t)} x_k(t)}{\sum_{t=1}^{m_{kj}} e^{-\alpha_x r_{ki}(t)}} \qquad z_2^{ki} = \frac{\sum_{t=1}^{m_{kj}} e^{-\alpha_y r_{ki}(t)} y_k(t)}{\sum_{t=1}^{m_{kj}} e^{-\alpha_y r_{ki}(t)}} \qquad (37)$$

The Cluster Reallocation Process: The cluster reallocation process consists on dislocating the cluster's position from the grid vertice to the obtained gravity-center. After, we redefine another window, calculate the new gravity-center, dislocate the cluster to this center, and son on.

The reallocation stops if the distance between the cluster location and the gravity-center becomes irrelevant. That is considered to occur when the sum of all cluster reallocations becomes less than a small number ε. In our tests the value was ε =0.0001.

Figure 24 displays the same clusters shown in Figure 23 and an example of a data distribution around them. For clusters N_2 and N_3, we draw the series of windows that appear after each iteration. The clusters, attained to the grid, are reallocated in direction to the real one. Because inside the window N_2 there is no data, it is eliminated as a valid cluster.

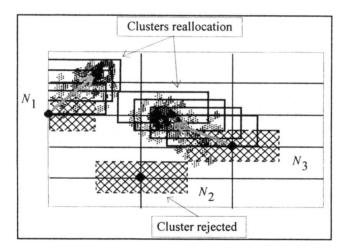

Fig. 24. Illustrative example of the cluster relacation process.

Figure 22 shows the clusters initially detected from the grid vertices for our illustrative example. Applying the reallocation process, Figure 25 shows the clusters, denoted by dark circles, being moved way from the grid in direction to the zones where the real clusters are localised, denoted with a white cross. The data linguistic structure is set through the membership functions obtained from the projection of the reallocated clusters in each coordinate x and y.

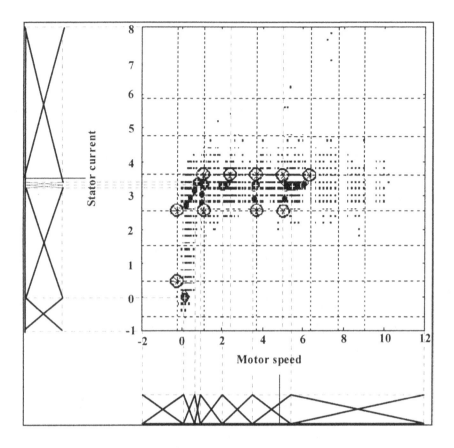

Fig. 25. Figure displaying the realocation of the initial clusters. After, their projection constructs the linguistic structure expressing the relations between the two variables.

5.2 Structure Identification Using the Autonomous Mountain-Clustering Algorithm

In this section, we describe the application of the autonomous mountain-clustering algorithm in identify the premise structure of the fuzzy rule-base. By premise structure, we mean the number of fuzzy sets and their position in each universe of discourse composing the rule antecedent part. The algorithm uses the learning set obtained from the electro-mechanical system described in section 2.

Using a "Trial-and-Error" Process for Fuzzy Structure Identification

Without applying any pre-processing algorithm, the electro-mechanical system is modelled using a "trial-and-error" process to identify the number of membership functions of each antecedent variable. The membership functions for all variables are considered displaced symmetrically by the universe of discourse, all functions with the same width and a symmetric triangular shape.

The best partition found is 11 fuzzy sets for each antecedent variable and 13 partitions for the consequent one. The fuzzy model obtains a quadratic error performance of $E=0,2539$. Figure 26 displays the error signal evolution with those partitions.

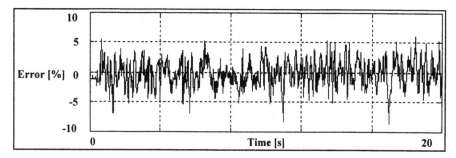

Fig. 26. Error signal evolution when the model structure uses a "trial-and-error" process

Using the Mountain Clustering Algorithm for Fuzzy Structure Identification

The algorithm is applied using an initial grid partition with 11 membership functions for each antecedent variable, as in the previous case. The partition makes a grid of 11 ×11×11 vertices in the domain with each one being a cluster candidate. The clustering algorithm selects a subset of 21 cluster candidates to represent the data.

Figure 27 displays a three-dimensional view of the systems domain. The figure shows the data and the identified clusters. The algorithm uses the parameters calculated by "trial-and-error": $\alpha=\beta=1,0$ and $\delta=10,0$. The parameters α and β are, for this test, equal in all directions. The parameter δ is the last density value valid to the cluster election.

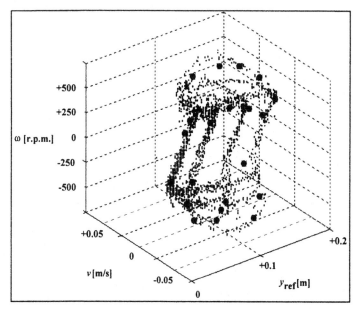

Fig. 27. Three-dimensional view of the data set with the identified clusters.

For a better clusters visualisation, Figure 27 is divided into three bi-dimensional figures (28, 29, and 30). Figures 28 and 29 present the distribution of the 9 membership functions attributed to variables y_{ref} and v. Figure 30 shows the 8 fuzzy partitions estimated for the third variable ω.

The fuzzy modeling algorithm is initialised attributing the estimated number and localisation of membership functions to each variable. The previous triangular functions are not necessarily distributed symmetrically as for the "trial-and-error" process. Comparing the results using the mountain-clustering method with the results using the "trial-and-error" process, we get a compressed and simplest fuzzy model (9-9-8 partitions). The partitions decreased about 52% maintaining the performance ($E = 0.2310$). Figure 31 shows the model error evolution, situated into 5%, for the new simplified fuzzy model

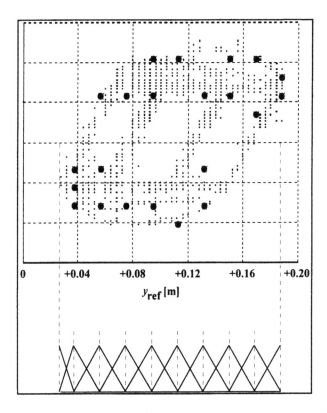

Fig. 28. A partition number of 9 membership functions attributed to the first variable (y_{ref}).

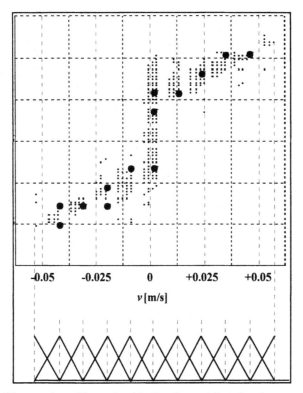

Fig. 29. A partition number of 9 membership functions attributed to the second variable (v).

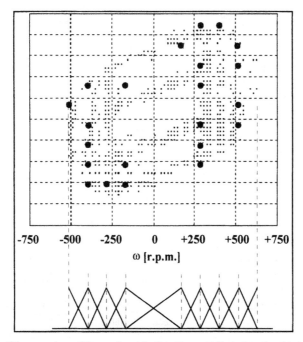

Fig. 30. A partition number of 8 membership functions attributed to the third variable (ω).

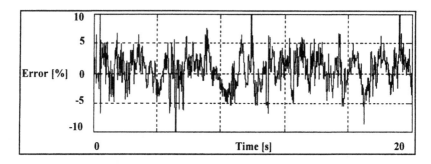

Fig. 31. Error evolution when applying the mountain-clustering method.

Using the Autonomous Algorithm Without Clusters Reallocation for Fuzzy Structure Identification

The objective of this test is verify if the parameter values, calculated automatically by the autonomous clustering algorithm, produce a fuzzy structure equivalent, or better, than the one obtained by the original mountain method. In this test, the reallocation process is not executed.

We use the same initial partition number of 11 fuzzy sets for each variable and the 21 identified clusters. The autonomous algorithm identifies 8 partitions for the first variable (y_{ref}), 5 partitions for the second one (v) and 7 for the third (ω). The extracted model presents a quadratic error of $E=0,2642$.

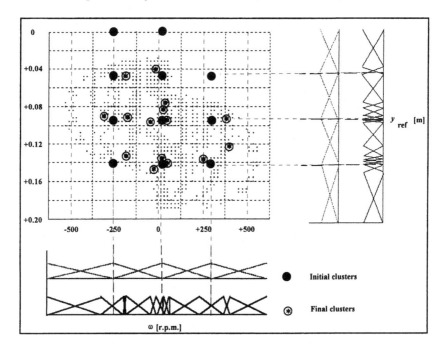

Fig. 32. Initial clusters (●)and the ones obtained before the adjustment process (⊛) for the first and second variable.

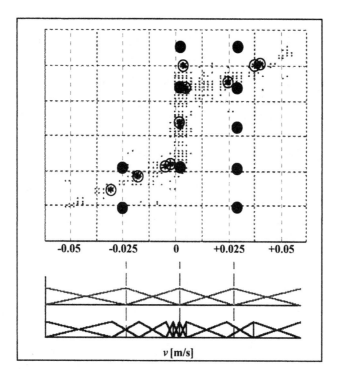

Fig. 33. Initial clusters (●) and the ones obtained before the adjustment process (⊛) for the third variable.

Figures 32-34 show that the final clusters, employing the parameters computed by the autonomous algorithm, are better distributed as before. The total fuzzy partitions decrease of 57% from that calculated with the mountain method (9-9-8), and maintains an analogue model performance.

Using the Autonomous Algorithm with Clusters Reallocation for Fuzzy Structure Identification

The main problem of the mountain clustering algorithm is setting the initial grid size. Since the cluster candidates are all fixed in the grid vertices within a symmetric structure, we need a lot of partitions to obtain accurate clusters. The idea of reallocating the final clusters is to break this fixed symmetric geometry, permitting to begin the algorithm with fewer partitions, decrease the computational cost, and generate a higher information quality.

The autonomous algorithm initialises the grid with 5 partitions for each variable and not 11 partitions as before. The computed clusters are reallocated and the initial symmetric structure desappears, as illustrated by Figures 32 and 33. These figures show the identified membership functions computed without cluster's reallocation. These functions are distributed symmetrically.

The initial clusters are represented by the symbol ● and the clusters, after the reallocation process, are displayed by the symbol ⊛. The final membership functions are displayed below the symmetric fuzzy sets.

The number of membership functions increases because, in reality, the fixed grid hides other clusters in parallel with the first ones. When the clusters are reallocated, the hidden clusters are moved away, and the information, which was hidden in the fixed grid, appears.

Testing the identified fuzzy structure in modeling the electro-mechanical system, the error results are presented in Figure 34. We obtain a similar model performance ($E = 0.2645$) as for the anterior results, but with a much smaller computational cost for the clusters identification algorithm. Although the final model had a great number of fuzzy sets attributed to each variable, the final membership functions indicate better the spatial data distribution in the domain, simplifying and increasing the adaptation process, if any are used to reallocate the membership functions.

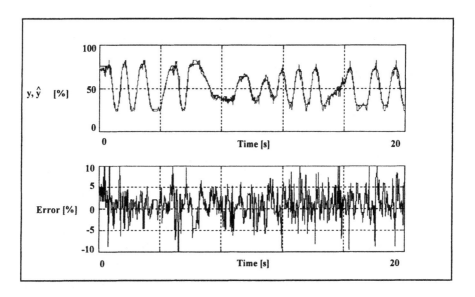

Fig. 34. Model and system response with the error signal when applying the autonomous mountain method using the cluster reallocation process.

6 Conclusion

In this paper, we presented two pre-processing methods for the identification of fuzzy systems structure. The first one was based on *Principal Component Analysis* and the other based on an approximate clustering algorithm called *Autonomous Mountain-Clustering Method*.

The approach using Principal Component Analysis is divided into two parts: the identification of the set containing the process dominant variables; and the application of a new system coordinates, composed by the principal components, in creating a new fuzzy rule-base. This makes the working domain more complete, reducing the fuzzy modeling errors.

The second approach is related with the number of membership functions attributed to each variable and their position in the universe of discourse. To find the optimal fuzzy partition of the input-space, the clustering algorithm provides us with a set of clusters describing the data set structure. For obtaining the membership functions, each cluster is projected onto all axes and localised in the respective cluster coordinate. The algorithm parameters are computed straight from data learning set without "trial-and-error" tentatives, and a clusters reallocation process is presented to eliminate the necessity of a greater number of partitions.

The results showed that using the autonomous algorithm with the cluster adjustment, the initial partition number is greatly reduced maintaining the same performance in the fuzzy model. With the adjustment, the clusters that were connected with the grid, can now be moved to more correct positions, increasing the information quantity (more fuzzy sets for each variable and better located) than that contained in the grid.

To investigate the potentialities of the proposed approaches on help modeling electro-mechanical systems with fuzzy-logic, an experimental electro-hydraulic system composed by a permanent-magnet synchronous motor driving a fixed displacement pump controlling a piston position was employed.

7 References

[1] R.R.Yager and D.P.Filev, "Approximate Clustering Via the Mountain Method", IEEE Trans. on Syst. Man and Cyb., vol. 24, No. 8, pp.1279-1284, August, 1994.

[2] P.J.Costa Branco, N.Lori and J.A.Dente, "An Autonomous Approach to the Mountain Clustering Method", The joint Third International Symposium on Uncertainty Modeling and Analysis and the Annual Conference of the North American Fuzzy Information Processing Society (ISUMA-NAFIPS'95), College Park, Maryland, USA, 1995.

[3] P.J.Costa Branco and J.A.Dente, "Intelligent Models for Electromechanical Systems", 6th European Conference on Power Electronics and Applications (EPE'95), Vol. 1, pp. 1148-1453, Sevilha, Spain, 1995.

[4] P.J.Costa Branco and J.A.Dente, "Automatic Modeling of an Electrical Drive System Using Fuzzy-logic". Proc. of First Int. Conf. of NAFIPS, IFIS, and NASA (NAFIPS/IFIS/NASA-94), pp. 441-443, Eds. Larry Hall et all., San Antonio - Texas, USA, IEEE Press, 1994.

[5] J.C.Bezdek and S.K.Pal, "Fuzzy Models for Pattern Recognition", Eds. New York: IEEE Press, 1992.

[6] M.Sugeno and T.Yasukawa, "A Fuzzy-Logic-Based Approach to Qualitative Modeling", IEEE Trans. on Fuzzy Systems, vol. 1, No. 1, pp. 7-31, February, 1993.

[7] Spiegel, R. Murray, "Schaum's Outline of Theory and Problems of Probability and Stastitics", Eds.Mcgraw-Hill, Brazil, 1977.

[8] Kleinbaum, Kupper and Muller, "Applied Regression Analysis and Other Multivariable Methods", 2nd. ed. Boston, Mass. :PWS-Kent, 1988.

A Generic Fuzzy Neuron and its Application to Motion Estimation

Abbas Z. Kouzani[*] and Abdesselam Bouzerdoum[**]

[*]School of Engineering
Flinders University
GPO Box 2100, Adelaide
SA 5001, Australia
E-mail: abbas@flinders.edu.au

[**]Dept. of Electrical and Electronic Eng.
The University of Adelaide
Adelaide
SA 5005, Australia
E-mail: bouzerda@eleceng.adelaide.edu.au

Abstract

The advantages of fuzzy sets and neural networks in emulating the human brain capabilities motivated the development of fuzzy neural networks. Various models of fuzzy neurons have been proposed as the basic element of fuzzy neural networks. In this paper, we introduce a generic fuzzy neuron as an extension of existing fuzzy neuron models. In our model, all the states of activity are given in terms of fuzzy sets with relative grades of membership distributed over the interval [0, 1]. The inputs and outputs are fuzzy sets over different universes of discourse. The connection, aggregation, and activation functions, which determine the operation of the neuron, are fuzzy relations. When the inputs to a function are fuzzy sets over the same universe of discourse, the function can be any fuzzy operation in class of triangular norms or triangular conorms. To evaluate the operation of the fuzzy neuron, a fuzzy neural network architecture based on the generic fuzzy neuron has been developed for motion estimation. The five-layer feedforward fuzzy neural network emulates a fuzzy motion estimation algorithm. Seven simplified versions of fuzzy neurons are defined and utilized in the fuzzy neural network. The results of simulations on thousands of 64×64, 6-bit synthetic image frames containing moving objects under different conditions are reported.

1 Introduction

Modelling various aspects of the human brain is a new field of artificial intelligence research concerned with the development of the next generation of intelligent systems. The human brain is superior to all kinds of modern computers in processing cognitive information, the information acquired by the peripheral nervous system. Whereas most of the traditional mathematical tools are based upon some absolute measures of information, the cognitive information is in the form of relative grades. Unlike the computational functions of traditional computers, the human brain acts upon the relative grades of raw information acquired by the neural sensory system.

To deal properly with uncertainties and imprecisions which arise from human thinking, mentation, cognition and perception, some special tools and techniques are required. In

1965, L.A. Zadeh published his paper on fuzzy sets as a means for representing uncertainty [22]. The type of uncertainty that this theory was meant to handle has as its roots the type of imprecision and ambiguity which is prevalent in human discourse and thought. The theory of fuzzy sets is based upon the notion of a relative graded membership.

Another difference between the human brain and conventional digital computers is its structure. It is believed that the brain consists of an enormous number of neurons highly interconnected by links with variable strengths, operating in parallel. Conventional computers, on the other hand, execute sets of instructions sequentially. To achieve the massively parallel distributed processing features of the brain, neural networks have been studied extensively since the work of McCulloch and Pitts, on neurons modelled as discrete decision-making elements, in 1943 [12]. In its simplest conception, the neural network may be described as a collection of neurons which interact among themselves through a highly interconnected synaptic network. The most striking aspect of such a network is the highly distributed manner in which information is stored and the high degree of parallelism by which it is processed. Moreover, like human brain, artificial neural networks have the ability to learn things.

To emulate the capability of the human brain in a machine, the attempt of utilizing fuzzy sets in the context of neural networks have begun after the inception of the fuzzy set theory and pursued from two different directions. Some researchers have utilized conventional neuron models to develop neural networks which are functionally equivalent to fuzzy inference systems [4][5][6][18]. These types of neural networks are trained using the learning rules which are mostly derived from the backpropagation algorithm. Other researchers, on the other hand, have developed neurons with fuzzy functions and fuzzy computations. They have replaced the conventional neurons with these fuzzy neurons in a neural network [1][2][8][9][11][16][17][20][21]. Most of those who have addressed learning algorithms, employed the learning rules which are mostly inspired by the backpropagation algorithm. This paper deals with the second group of fuzzy neural systems.

In the next section, a generic model of a fuzzy neuron is introduced as the basic element of our fuzzy neural system. The generic fuzzy neuron is a generalization of the existing models of fuzzy neurons. In Section 3, a fuzzy algorithm for motion estimation in presented. Then, a five-layer fuzzy neural network which implements the fuzzy algorithm is proposed. We define seven simplified types of fuzzy neurons to be employed in different layers of the fuzzy neural network architecture. The simulation results are presented and discussed in Section 4, which is followed by some concluding remarks.

2 The Generic Fuzzy Neuron

In this section, a generic fuzzy neuron is introduced to form the basic computational element of a fuzzy neural network that will be discussed later. The generic fuzzy neuron model is a generalization of existing fuzzy neuron models. It is inspired by the Gupta and Knopf's fuzzy neuron [1]; however, it differs from that of Gupta and Knopf and other models in the following ways:

a) All the variables involved in the generic fuzzy neuron are allowed to be fuzzy sets over different universes of discourse.

b) All the functions that specify the characteristics of the neuron are chosen to be fuzzy relations (when the universes of discourse are similar, the fuzzy relation can be replaced by any fuzzy operation).

c) The output of each function, which is a fuzzy set, is obtained using the *-* composition of inputs to the function and the corresponding fuzzy relation.

d) Each fuzzy neuron can be simplified to represent an entire fuzzy inference rule with any number of propositions.

These differences make it possible to carry out any fuzzy computation on the input data and express any kind of ambiguous relationship. This leads to the following definition:

Definition: A *generic fuzzy neuron* (Figure 1) consists of connection functions, an aggregation function and an activation function. It has N inputs $I_1, I_2, ..., I_N$, each input consisting of a finite set of elements.

Let
$$X_1 = \{x_{11}, x_{12}, ..., x_{1n_1}\}, \qquad X_2 = \{x_{21}, x_{22}, ..., x_{2n_2}\},...,$$
$X_N = \{x_{N1}, x_{N2}, ..., x_{Nn_N}\}$ denote the N finite sets of inputs. The inputs $I_1, I_2, ..., I_N$ are fuzzy sets in the universes of discourse $X_1, X_2, ..., X_N$, characterized by membership functions $\mu_{I_1}, \mu_{I_2}, ..., \mu_{I_N}$. The inputs are weighted with $W_1, W_2, ..., W_N$, which are fuzzy sets in the universes of discourse $S_1 = \{s_{11}, s_{12}, ..., s_{1i_1}\}$, $S_2 = \{s_{21}, s_{22}, ..., s_{2i_2}\}, ..., S_N = \{s_{N1}, s_{N2}, ..., s_{Ni_N}\}$. The weighting operation is done through *connection functions* of the form:

$$A_i = c_i(I_i, W_i) \tag{1}$$

Let C_i be a 3-ary fuzzy relation in $[X_i] \times [S_i] \times [U_i]$, the output of a connection function is expressed by the *-* composition of X_i, W_i, and C_i which is the fuzzy set A_i in the universe of discourse $U_i = \{u_{i1}, u_{i2}, ..., u_{ij_i}\}$. The connection outputs, A_i's, may be classified as either excitatory or inhibitory. Let B_i's be the direct inputs to the neuron, where

$$B_i = \begin{cases} A_i & \text{an excitatory input} \\ \overline{A}_i & \text{an inhibitory input} \end{cases} \tag{2}$$

$\overline{A}_i$ denotes complement of the fuzzy set A_i, which is defined by the following membership function

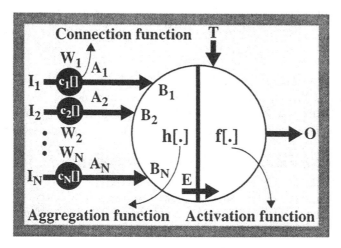

Figure 1: The generic fuzzy neuron.

$$\mu_{\overline{A}_i}(x) = 1 - \mu_{A_i}(x) \tag{3}$$

The set B_i is a fuzzy set in the same universe, U_i, as A_i.

In the next stage, an *aggregation function* $h[.]$ provides an input to the last stage as follows

$$E = h(B_1, B_2, ..., B_N) \tag{4}$$

Let H be an $(N+1)$-ary fuzzy relation in $[U_1] \times [U_2] \times ... \times [U_N] \times [V]$, the output of the aggregation function is expressed by the *-* composition of $B_1, B_2, ..., B_N$, and H, which is the fuzzy set E in the universe of discourse $V = \{v_1, v_2, ..., v_k\}$.

The output of the fuzzy neuron is determined by an *activation function*, in the last stage, as follows

$$O = f(E, T) \tag{5}$$

where f is the activation function and T is the threshold input which is a fuzzy set over the universe of discourse $Z = \{z_1, z_2, ..., z_l\}$. Let F be a 3-ary fuzzy relation in $[V] \times [Z] \times [Y]$. The output O is determined by *-* composition of E, T, and F, and is a fuzzy set in the universe of discourse $Y = \{y_1, y_2, ..., y_m\}$.

The generic fuzzy neuron can be simplified if some of the fuzzy sets I_i, W_i, A_i, B_i, E, T, and O are defined over the same universe of discourse, e.g. $X = \{x_1, x_2, ..., x_n\}$. In this case, the related fuzzy relations C_i, H, or F can be replaced with any operator from the two basic classes of operations on fuzzy sets, which are the triangular norms (t-norms) and the triangular conorms (t-conorms or s-norms), such as min, max, bounded-difference, algebraic product, and so on. Many types of fuzzy neurons can be defined by changing the functions $c[.]$, $h[.]$, and $f[.]$. In the next section seven types of fuzzy neurons utilized in construction of a fuzzy neural network for motion estimation are defined.

In general, the weights, the connection functions, the aggregation function, the threshold, and the activation function could be tuned during the learning phase. Therefore, the fuzzy neural network which will be constructed with neurons of this type can learn from experience. Although in this paper no learning rule was used.

3 A Fuzzy Neural Network for Motion Estimation

In this section, we introduce an architecture of a fuzzy neural network designed for detection of moving objects and estimation of their velocity. But before describing the architecture of the network, let us first review the algorithm we proposed for motion estimation [7].

3.1 The Fuzzy Motion Estimation Algorithm

Motion can be regarded as orientation in the spatio-temporal domain [3]. This fundamental fact has formed the basis of the algorithm designed for motion estimation in this work. In the proposed method, motion evaluation does not depend on object detection. Calculation of the velocity vector, consisting of average speed and motion direction, is done for each pixel individually. Here, we assume that the temporal sampling interval is very short and that no significant changes occur between two consecutive frames. Moreover, each pixel in an image can move to one of its nine neighbouring pixels in the successive frame. This simplifying assumption is made so as to reduce the complexity of the algorithm and simplify the fuzzy-neural network architecture. However, the assumption made is not a limitation of the approach, but a pratical simplification. The approach can easily be extended to deal with the objects travelling more than one pixel per frame.

It is well known that the computation of optical motion estimation is an ill-posed, ill-conditioned problem [13]. When noise is present in the input image, or when pixels have the same intensities, it is virtually impossible to find out which of the pixels in one frame has moved to a given pixel in the next frame if the matching criterion is based on pixel intensity values only. A solution to this problem is to assign a sector of the image consisting of the given pixel and its neighbouring pixels as a representative for the given pixel, then compare the sector from the second image with sectors of equal size from the first image within a certain search area. In our approach, the representative sector of a pixel is selected to be the 3×3 region comprising the pixel itself and its eight neighbouring pixels, as shown in Figure 2.

The search area is limited to the 3×3-neighbouring sectors. To find out which of the nine possible pixels in the first image has moved to a given pixel in the second image, we search for the maximum similarity between the representative sector of the given pixel and the sectors representing its neighbouring pixels, in the first image. This is done using fuzzy relations and fuzzy operations.

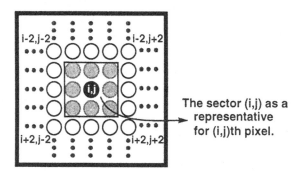

i-2,j-2 i-2,j+2

i+2,j-2 i+2,j+2

The sector (i,j) as a representative for (i,j)th pixel.

Figure 2: Representative sector of the (i,j)th pixel.

Simple template matching or correlation schemes suffer from the problem of ambiguity in matching. In other words, if there exist more than one possible match, these methods cannot extract the right match [19]. In our approach, a fuzzy control mechanism is employed to strengthen or weaken the results of the comparisons. This allows the history of motion to be taken into account, which resolves the ambiguity problem in simple template matching or correlation schemes. Obviously, it is feasible to extend the sector size and/or the search area at the expense of computational complexity and cost.

A gray-tone image possesses ambiguity within each pixel because of the possible multivalued levels of brightness a pixel can have [14]. If the gray levels are scaled to lie in the range [0, 1], we can view a gray image as a fuzzy set. With the concept of fuzzy sets, an image X of size $M \times N$ and L gray-levels can be considered as an array of fuzzy singletons, each having a value of membership denoting its degree of brightness relative to some brightness level l, $l = 0, 1, ..., L - 1$. In the notation of fuzzy sets, X may be defined as

$$X = \bigcup_m \bigcup_n \mu_{mn}/x_{mn} \qquad (6)$$

$$m = 1, 2, ...M, n = 1, 2, ...N$$

where μ_{mn}/x_{mn} $(0 \le \mu_{mn} \le 1)$ is a fuzzy singleton which denotes the grade of possessing some brightness property μ_{mn} by the (m, n)th pixel [15]. The fuzzy property μ_{mn} may be defined in a number of ways with respect to any brightness level. In this work, a second-order S function [10] has been used as follows

$$S(x; a, b, c) = \begin{cases} 0 & \text{if } x \le a \\ S_1 & \text{if } a < x \le b \\ S_2 & \text{if } b < x \le c \\ 1 & \text{if } x > c \end{cases} \qquad (7)$$

where

$$S_1(x;a, b, c) = \frac{(x-a)^2}{(b-a)(c-a)} \qquad (8)$$

and

$$S_2(x;a, b, c) = 1 - \frac{(x-c)^2}{(c-b)(c-a)} \qquad (9)$$

Before proceeding with the details of how the similarity between a given sector in the second image and its nine neighbouring sectors in the first image is computed, let us describe the parameters we employ to measure the similarity and the variables we use to store the average velocity and motion direction of a given pixel (see Figure 3).

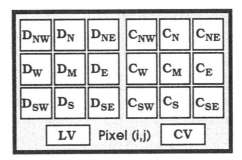

Figure 3: Variables assigned to each pixel.

Last Velocity

LV is a six-bit register which contains the time interval between two consecutive movements of a given pixel. Its content is updated whenever a new movement to that pixel is detected and used to calculate the average speed.

$$\text{Speed} = 1 - LV/63 \qquad (10)$$

where speed is measured in *pixels per frame* (ppf); the maximum speed that can be measured is 1 ppf, as stated in the assumption earlier. Note that the direction of movement is determined using the direction variables defined below. Since we are using six bits for LV, the pixel's speed can take only one of 64 possible values. This choice is arbitrary. Increasing the register length will increase the number of speed values that can be computed, but this also increases the complexity and size of the fuzzy neural network.

Current Velocity

CV is a seven-bit counter which represents the current velocity being measured for a given pixel. The content of this variable, which is increased by one at each sampling instant, measures the number of frames (sampling times) taken between two consecutive movements. When a new movement to a pixel is detected, the content of its CV variable is copied to the LV register, and then CV is reset to zero. Note that as soon as CV exceeds 63, its content is clipped to 63 and copied to LV. This means that no movement has happened during the last 63 input frames. Although seven bits are used for CV, only the first six bits are used in the calculation of the current velocity. The purpose of adding an extra bit to CV is to establish the updating mechanism of the control variables which will be described later (Equation (12), Figure 6).

Direction Variables

The variables $D_W, D_{NW}, D_N, D_{NE}, D_E, D_{SE}, D_S, D_{SW}$, and D_M are nine 6-bit registers called the direction variables. Each direction variable denotes a membership degree of possessing the velocity value stored in LV by the pixel in the corresponding direction, e.g. north east, middle, south west, etc. These variables are updated together with LV when a new movement to the pixel is detected. Calculation of the direction variables is performed using the following function

$$y = \exp\left(-\beta^2 n^2\right) \tag{11}$$

where β is a constant parameter, which can be selected by trial and error or by a learning algorithm, and n is an integer in the interval $[0, 5]$.

The values of the integer n used to update the direction variables in Equation (11) are determined based on the estimated direction of motion. The pixel that moved is given $n = 5$, its two nearest neighbours are given $n = 4$, and so on. Figure 4 illustrates two possible motion directions and the values of n corresponding to each direction variable. Note that n is always equal to 2 for the middle variable D_M.

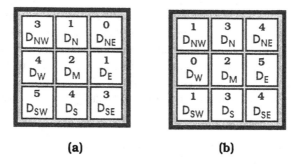

(a)   (b)

Figure 4: Numbering example of the direction variables. (a) Movement from the south-west pixel (↗). (b) Movement from the east pixel (←).

Control Parameters

C_W, C_{NW}, C_N, C_{NE}, C_E, C_{SE}, C_S, C_{SW}, and C_M are nine 6-bit registers called control parameters. These parameters are used in the calculation of the similarity measure. They are updated once each sampling interval. In the update process, the values of LV and CV, and the value of the corresponding direction variable are used to calculate the new value of each control parameter. Figure 5 shows the variables that are needed in the calculation of the north-west control parameter of a given pixel.
The update mechanism is formulated as follows:

$$C_X = \begin{cases} \dfrac{D_X}{LV} \cdot CV & \text{if } 0 \le CV \le LV \\[2mm] D_X\left(1 - \dfrac{CV - LV}{64}\right) & \text{if } LV < CV \le LV + 63 \\[2mm] 0 & \text{if } LV + 63 < CV \end{cases} \qquad (12)$$

where $0 \le D_X \le 63$, $0 < LV \le 63$, and $0 \le CV \le 127$.

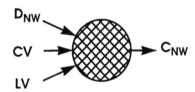

Figure 5: The variables needed for the calculation of the NW control parameter C_{NW}.

Figure 6 illustrates a graphical representation of the operation described by Equation (12) for updating the control parameters. The control parameters are used to suppress the effect of input noise and solve the *correspondence problem*. That is, even when there is more than one good match between a sector in the second image and its neighbouring sectors in the first image, the best match can still be found. The reason is that both time domain and spatial domain information is included in the calculation of the best-matched sector. In the spatial domain the similarity is found using the brightness patterns from two consecutive frames, whereas in the time domain the results of all previous calculations of similarity are used to strengthen or weaken the similarity between sectors.

As shown in Figure 6, a control variable is a dynamic parameter whose value is changed at each sampling instant. Since the sampling interval is small enough, an object can only change speed and direction smoothly and gradually. When a movement to a given pixel occurs, the movement will be transferred to the next pixel with approximately the same speed and same direction. Therefore, the possibility of transferring the motion from the receiving pixel to the next pixel is highest when the time index kept in the CV counter of the receiving pixel is nearly equal to the content of the *LV* register. This fuzzy mechanism reinforces the similarity measure in the next pixel at the appropriate time and weakens it at other times. The maximum reinforcement is achieved in the direction in which D_X has its maximum value specified by Equation (11).

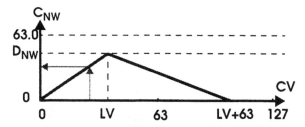

Figure 6: Graphical representation of a control parameter updating operation.

Having explained how to extract the control parameters, we now describe how the similarity measure is calculated. First, all the similarities between the representative sector of a given pixel in the second image and its nine neighbouring sectors in the first image are measured. The operation on the sector's brightness patterns is done using the absolute-difference fuzzy operator [7] . As a result, nine similarity values are obtained for each pixel. Then the nine measured similarities are strengthened by nine control parameters in each pixel. The control parameters which are used for this purpose come from the neighbouring pixels, except the middle control parameter which is taken from the pixel itself.

The strengthening operation is performed using the algebraic product fuzzy operator [7]. Nine match indicators are calculated for each pixel as shown in Figure 7. Each match indicator expresses the final degree of similarity between the representative sector of a pixel and its nine neighbouring sectors in the preceeding image. The best match sector can then be easily specified using the max-bounded-difference compositional rule of inference [7].

After the bes match sector is extracted for a given pixel, its average velocity and movement direction are readily available. If the best match to the sector representing the (i, j)th pixel is the (i, j)th sector in the previous frame, then no movement has occurred at this pixel. In this case, the *CV* counter is increased. If the content of the *CV* counter exceeds 63, there is no movement to this pixel. If the best match is not the (i, j)th sector, then a movement has occurred from the pixel with the highest best match index to the given pixel. In this case, the content of *CV* which represents the time of travel is transferred to *LV* . The *CV* counter is then reset to zero.

The features of the proposed algorithm are:

1) Suppressing the effect of noise.

2) Solving the correspondence problem, i.e., the problem of finding the correct match among other possible matches in brightness patterns.

3) Robustness against changes in luminance.

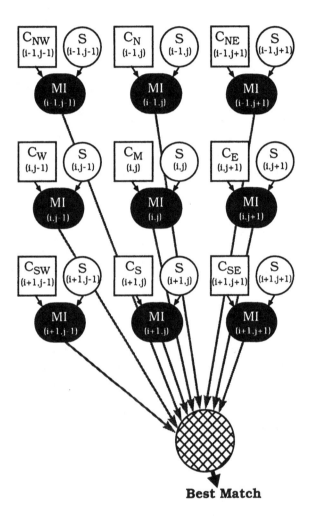

Best Match

Figure 7: The parameters used for extracting the best match between the representative sector of the (i,j)th pixel and its nine neighbouring sectors.

3.2 Architecture of the Fuzzy Neural Network

The proposed fuzzy neural network for motion estimation is a five-layer feedforward network with a hierarchical structure as shown in Figure 8. The inputs to the network are two matrices of 64×64 pixels of 64 gray-levels. The outputs are velocity vectors consisting of speed and motion direction for each individual pixel of the image. In the following we describe each layer of the fuzzy neural network individually.

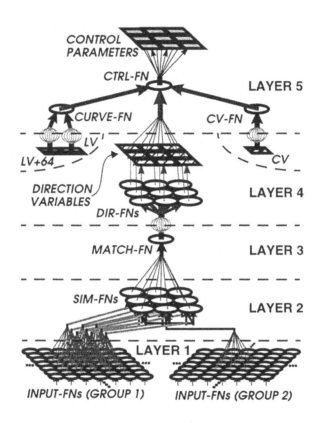

Figure 8: Architecture of the fuzzy neural network for motion estimation.

First layer

The first layer is the input layer which accepts a pattern into the network. It consists of two sets of 64×64 input fuzzy neurons (INPUT-FNs) shown in Figure 9. The first set of INPUT-FNs is allocated to the current frame. The second set is for the previous frame. There is a 64×64 memory unit reserved for storing the previous frame. In each sampling interval, when a new image frame is acquired, the current image is overwritten into the memory forming the previous image frame.

Each INPUT-FN in this layer corresponds to one-pixel of the previous or current frame. To express the input image in terms of fuzzy sets, we consider the image as an array of fuzzy singletons, each having a membership value denoting its degree of brightness in the interval [0,1] (Equation (6)). To determine the membership value, μ_{mn}, we use the second order S function as follow:

$$\mu(x) = S(x;0, 31, 63) \qquad (13)$$

where x is the pixel intensity in the interval [0, 63]. Each INPUT-FN determines a membership value for its input pixel by implementing the function described by Equation (13). The INPUT-FN has one input which is a fuzzy set over a universe of discourse, X, with 64 elements. All the elements are connected to the given input pixel. The weight is a fuzzy set over the same universe of discourse whose elements are set as follows:

$$W = \{\frac{0}{63}, \frac{1}{63}, ..., \frac{63}{63}\} \qquad (14)$$

Since the input and the weight are fuzzy sets over the same universe of discourse, we employ the absolute difference fuzzy operator, $\boxminus$, as connection functions. The input A_1 to the neuron is excitatory, therefore

$$B_1 = A_1 \qquad (15)$$

The aggregation function $h[.]$ is a 2-ary fuzzy relation, H, which is defined by a 64×64 matrix as follows:

$$H = \begin{bmatrix} 1 & 0 & 0 & ... & 0 \\ 0 & 1 & 0 & ... & 0 \\ 0 & 0 & 1 & ... & 0 \\ ... & ... & ... & ... & 0 \\ 0 & 0 & 0 & ... & 1 \end{bmatrix} \qquad (16)$$

The output of the aggregation function is the fuzzy set E determined by the fuzzy max-min composition of B_1 and H. The fuzzy set E will have 63 elements whose values are 1, and one element whose value is 0. The intensity of the input pixel specifies which elements should be zero. The INPUT-FN has no threshold. The output of INPUT-FN is determined by the activation function $f[.]$ which is a 2-ary fuzzy relation, F, defined by the following 64×1 matrix:

$$F = \begin{bmatrix} S(0;0, 31, 63) \\ S(1;0, 31, 63) \\ ... \\ S(63;0, 31, 63) \end{bmatrix} \qquad (17)$$

The output O, is a fuzzy set in the universe of discourse Y with only one element. It is determined by the fuzzy min-max composition of E and F; it expresses the membership degree of the input brightness in the interval [0,1].

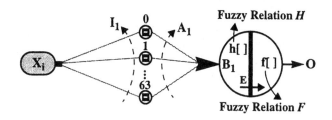

Figure 9: Input fuzzy neuron (INPUT-FN).

Second layer

The purpose of the second layer is to determine the similarity values between the representative sector of each pixel in the current frame and its neighbouring sectors in the previous frame. The absolute difference fuzzy operator, $\boxminus$, is employed for this purpose. There are nine similarity fuzzy neurons (SIM-FNs) in this layer for each pixel. For the 64×64-pixel input images, a total number of $9 \times 64 \times 64$ SIM-FNs are allocated to the second layer. We choose a 3×3 universe of discourse containing 9 elements (pixels), $X = \{x_{11}, x_{12}, x_{13}, x_{21}, x_{22}, x_{23}, x_{31}, x_{32}, x_{33}\}$, to express the representative sector of each pixel as described before. For a given pixel, the single output of its corresponding INPUT-FN plus eight single outputs of its neighbouring pixels determine the nine elements of the universe of discourse, i.e., the pixel's representative sector.

As illustrated in Figure 10, each SIM-FN has two inputs, I_1 and I_2, which are fuzzy sets over the same universe of discourse, X. The input I_1 is connected to the outputs of the corresponding INPUT-FNs in the previous frame, whereas, the input I_2 is fed by the related current frame INPUT-FN outputs. There is no weighting operation for both I_1 and I_2, i.e. the connection functions are null functions. The input to the neurons are both excitatory thus we have:

$$B_1 = A_1 = I_1 \quad \text{and} \quad B_2 = A_2 = I_2 \tag{18}$$

Since both inputs are fuzzy sets over the same universe of discourse, we employ the absolute difference fuzzy operator as the aggregation function of SIM-FNs. The result is a fuzzy set E in the same universe of discourse. E has nine elements each denoting the absolute difference between the corresponding input elements.

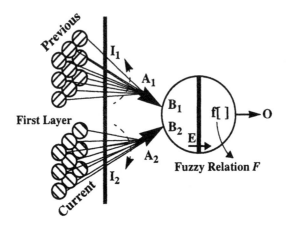

Figure 10: Similarity fuzzy neuron (SIM-FN).

The output of SIM-FN is determined by the neuron's activation function. The employed activation function is a 2-ary fuzzy relation, F, defined by the following 9×1 matrix:

$$F = \begin{bmatrix} 1/9 \\ 1/9 \\ \dots \\ 1/9 \end{bmatrix} \tag{19}$$

The output O, is a fuzzy set in the universe of discourse Y with only one element. It denotes bounded sum of the nine elements of the fuzzy set E. This is done with the aid of bounded-sum-algebraic-product composition of E and F. The single output expresses the similarity value between its two input sectors in the interval [0, 1]. When two sectors are exactly the same, O will be equal to one.

Third layer

The third layer is to determine the match indexes between the representative sector of a pixel in the current frame and its neighbouring sectors in the previous frame. This is done using similarity values and control parameters described before. For each pixel, the nine single-element similarities measured by nine SIM-FNs in the second layer are strengthened with nine control parameters. This operation is carried out through a match fuzzy neuron (MATCH-FN). There are 64×64 MATCH-FNs in the third layer each neuron is allocated to a given pixel. We choose a 3×3 universe of discourse X in the input, containing 9 elements. For a given pixel, the nine SIM-FNs single outputs form the nine elements of the universe of discourse.

As demonstrated in Figure 11, a MATCH-FN has only one input, I_1. It is a fuzzy set in the universe of discourse, X. Each element of the fuzzy set I_1 is connected to the corresponding single output of a SIM-FN in the second layer. I_1 is strengthened through connection functions using weights. The weight is a fuzzy set over the universe of discourse X and its elements are set as follows:

$$W = \{ c_{SE_{(i-1,j-1)}}, c_{S_{(i-1,j)}}, c_{SW_{(i-1,j+1)}}, c_{E_{(i,j-1)}},$$
$$c_{M_{(i,j)}}, c_{W_{(i,j+1)}}, c_{NE_{(i+1,j-1)}}, c_{N_{(i+1,j)}}, c_{NW_{(i+1,j+1)}} \} \tag{20}$$

The control parameters and therefore the weight W are changed in each sampling interval. Since the input and weight are fuzzy set over the same universe of discourse, we employ the algebraic product fuzzy operator as connection function. The output of the connection function is a fuzzy set A_1 treated as an excitatory input to the fuzzy neuron; therefore,

$$B_1 = A_1 \tag{21}$$

The aggregation function is a 2-ary fuzzy relation, H, which is defined by a 9×1 matrix as follows

$$H = \begin{bmatrix} 1 \\ 1 \\ 1 \\ \dots \\ 1 \end{bmatrix} \tag{22}$$

The output of the aggregation function is a fuzzy set E determined by the fuzzy max-min composition of B_1 and H. The fuzzy set E will have only one element whose value denotes the maximum value of the nine elements of the fuzzy set B_1.

The threshold input T in MATCH-FN is a fuzzy set in the universe of discourse X as the input. It has nine elements, each element is connected to the corresponding element of fuzzy set A_1. Therefore we have

$$T = A_1 \tag{23}$$

This arrangement is to determine the difference between each match indicator, which is an element of the fuzzy set A_1, with the maximum value of the nine match indicators. The output of MATCH-FN is determined by the activation function which is a 3-ary fuzzy relation, F, defined by the following 9×9 matrix

$$F = \begin{bmatrix} 0 & 1 & 1 & \dots & 1 \\ 1 & 0 & 1 & \dots & 1 \\ 1 & 1 & 0 & \dots & 1 \\ \dots & \dots & \dots & \dots & 1 \\ 1 & 1 & 1 & \dots & 0 \end{bmatrix} \tag{24}$$

The output O, is a fuzzy set in the universe of discourse X with nine elements. It is determined by the fuzzy max-bounded-difference composition of E, T, and F. Each element of the output fuzzy set denotes the difference between the corresponding element of the input to the neuron and the element that carries the maximum value.

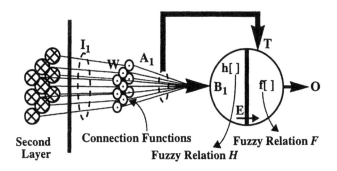

Figure 11: Match fuzzy neuron (MATCH-FN).

Fourth layer

The calculation of the direction variables, D_X, is carried out in the fourth layer. This operation is carried out through nine direction fuzzy neurons (DIR-FNs). There are $9 \times 64 \times 64$ DIR-FNs in the forth layer, each nine neurons are allocated to a given pixel. We choose a 3×3 universe of discourse X for the input, containing nine elements.

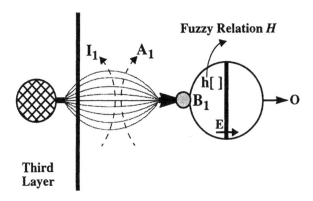

Figure 12: Direction fuzzy neuron (DIR-FN).

As shown in Figure 12, each DIR-FN has only one input, I_1. It is a fuzzy set in the universe of discourse, X. The connection between the Layer 3 outputs and the Layer 4 inputs is defined as follows:

$$I_{11} = I_{12} = \ldots = I_{19} = O \tag{25}$$

There is no weighting operation for I_1, i.e. the connection functions are null functions. The output of a MATCH-FN is complemented by setting the input to DIR-FN as inhibitory, thus higher values in the input elements indicate higher match degree from the third layer process; therefore,

$$B_1 = \overline{A_1} \tag{26}$$

The aggregation function is a 2-ary fuzzy relation, H, which is defined by a 1×9 fuzzy matrix. The contents of the fuzzy matrices are different for the nine DIR-FNs. The nine fuzzy matrices of the nine DIR-FNs, for a given pixel in the fourth layer, are defined in Appendix B. The output of the aggregation function is determined by the fuzzy max-algebraic-product composition of B_1 and H, which is a fuzzy set E in the universe of discourse V containing only one element. E determines a value for the corresponding direction variable of a pixel using $\exp\left(-\beta^2 x^2\right)$. As can be seen from the contents of the matrices, in a given DIR-FN, this arrangement allows the neuron to behave like a lens so that the neuron focuses on one of the input elements which is directly related to the direction value controlled by the neuron. However, the neuron uses the other elements in the computation of the direction value with lower degree of importance. β is assigned to 0.5 in the system simulations, based on a few experiments which were conducted to obtain a suitable value for β; however, the best value for this parameter should be extracted from a learning process. We will introduce a learning algorithm for this purpose in our future work.

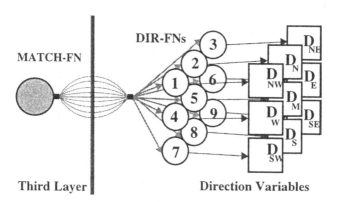

Figure 13: The connections between DIR-FNs and direction variables.

There is no threshold input and no activation function for DIR-FN, i.e., the neuron's output is equal to the aggregation function output

$$O = E \tag{27}$$

The output O has one element that denotes a value for the corresponding direction variable of a pixel in the interval [0, 1]. For a given pixel, Figure 13 shows how each direction variable is driven by a DIR-FN. The direction variables which are memory cells accept the DIR-FN outputs when a movement to a pixel is detected, i.e., the output of the fifth neuron is not the maximum one.

Fifth layer

The fifth layer is the last layer of the fuzzy neural network. The calculation and updating of the control parameters are carried out in this layer. There are three different types of fuzzy neurons used for extracting the control parameters of a given pixel. Therefore, the total number of the neurons in Layer 5 are $3 \times 64 \times 64$. For a given pixel, the inputs to this layer are the pixel's LV register, the pixel's CV counter, and nine direction parameters. The output of this layer are nine control parameters whose values vary in the interval [0, 1]. As mentioned in the previous section, the calculation of the control parameters is done using Equation (8). This function is implemented by a curve fuzzy neuron (CURVE-FN), a current velocity fuzzy neuron (CV-FN), and a control fuzzy neuron (CTRL-FN).

A CURVE-FN builds the curve shown in Figure 6 using the LV value. It has two inputs I_1 and I_2 in the same universe of discourse X with 128 elements. All the elements of the first input are connected to LV and the second input to $LV + 64$. Since LV is a binary number between 1 and 63, we can simply add an extra bit which is always "high" as the MSB. Therefore, we will have $LV + 64$ as the result. There are two weights W_1 and W_2 which are fuzzy sets over the same universe of discourse, X. W_1 and W_2 elements are set as follows:

$$W_1 = W_2 = \{ \frac{0}{127}, \frac{1}{127}, ..., \frac{127}{127} \} \tag{28}$$

We employ the algebraic-divide fuzzy operator (see Appendix A), as connection functions. The input A_1 to the neuron is excitatory forming the first part of the curve, interval [0, LV], whereas the input A_2 is inhibitory forming the second part of the curve, interval [LV, $LV+63$]; therefore,

$$B_1 = A_1 \ , \ B_2 = \overline{A_2} \tag{29}$$

The fuzzy set B_1 forms the left part of the curve and the fuzzy set B_2 forms its right part. The aggregation function is a min fuzzy operator. The output of the aggregation function is a fuzzy set E over the same universe of discourse X with 128 elements. It contains 128 elements of the whole curve. There is no threshold input and no activation function for the CURVE-FN. The output O of the neuron is defined as follows:

$$O = E \tag{30}$$

It represents the curve shown in Figure 6 without considering the scaling effect of the corresponding direction variable. This curve will be scaled down in the CTRL-FN to achieve the curve described by Equation (12).

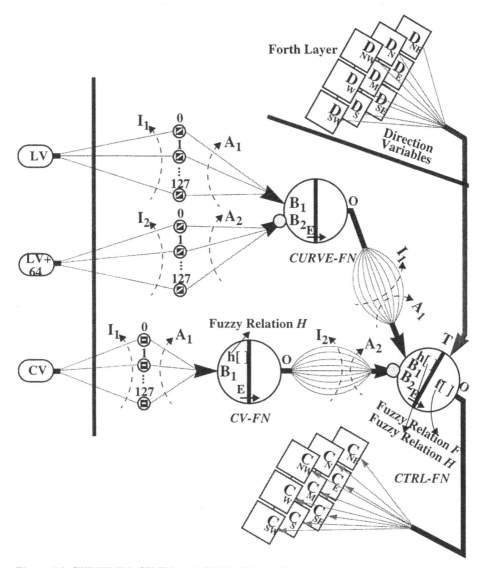

Figure 14: CURVE-FN, CV-FN, and CTRL-FN plus their connections.

The CV-FN is similar to the INPUT-FN but it does not employ an activation function. Each CV-FN has one input which is a fuzzy set over the universe of discourse X with 128 elements. All the elements are connected to the CV's output. The weight is a fuzzy set over the same universe of discourse X and its elements are set out as follows:

$$W = \{\frac{0}{127}, \frac{1}{127}, ..., \frac{127}{127}\} \tag{31}$$

We employ the absolute difference fuzzy operator (see Appendix A), as connection functions. The input A_1 to the neuron is excitatory; therefore,

$$B_1 = A_1 \tag{32}$$

The aggregation function $h[.]$ is a 2-ary fuzzy relation, H, which is defined by a 128×128 matrix as follows:

$$H = \begin{bmatrix} 1 & 0 & 0 & ... & 0 \\ 0 & 1 & 0 & ... & 0 \\ 0 & 0 & 1 & ... & 0 \\ ... & ... & ... & ... & 0 \\ 0 & 0 & 0 & ... & 1 \end{bmatrix} \tag{33}$$

The output of the aggregation function is a fuzzy set E determined by the fuzzy max-min composition of B_1 and H. The fuzzy set E will have 127 elements whose values are 1, and one element whose value is 0. The CV-FN has no threshold input and no activation function; output is defined as

$$O = E \tag{34}$$

It is a fuzzy set with 128 elements which denotes the input in terms of fuzzy sets.

A CTRL-FN has two inputs I_1 and I_2 in the same universe of discourse X with 128 elements. The elements of the first and the second inputs are connected to the corresponding elements of the CURVE-FNs output and the CV-FNs output respectively. There are no connection functions and the inputs are excitatory and inhibitory, respectively. We have

$$B_1 = A_1 \quad B_2 = \overline{A_2} \tag{35}$$

The input A_2 to the neuron is set to be inhibitory for complementing the output of CV-FN; therfore, the fuzzy set B_2 will have 127 elements whose values are 0, and one element whose value is 1. For example, if $CV = 45$, the 45th elemnt has will be 1.

The aggregation function is a 3-ary fuzzy relation, H, which is defined by a $128 \times 128 \times 1$ matrix as follows:

$$H = \begin{bmatrix} 1 & 0 & 0 & ... & 0 \\ 0 & 1 & 0 & ... & 0 \\ 0 & 0 & 1 & ... & 0 \\ ... & ... & ... & ... & 0 \\ 0 & 0 & 0 & ... & 1 \end{bmatrix} \tag{36}$$

The output of the aggregation function is a fuzzy set E determined by the fuzzy max-min composition of B_1, B_2, and H. The fuzzy set E will only have one element whose value denotes the output of the function described by Equation (12) to its input which is driven by CV. This value is scaled down by nine direction variables producing the new values of nine control parameters; this is done by the activation function. The threshold input T is a fuzzy set in the universe of discourse Z with nine elements. Each element of the threshold input is connected to one direction variable. The activation function is a 3-ary fuzzy relation, F, defined by the following $1 \times 9 \times 9$ matrix:

$$F = \begin{bmatrix} 1 & 0 & 0 & ... & 0 \\ 0 & 1 & 0 & ... & 0 \\ 0 & 0 & 1 & ... & 0 \\ ... & ... & ... & ... & 0 \\ 0 & 0 & 0 & ... & 1 \end{bmatrix} \tag{37}$$

The output O, is a fuzzy set in the universe of discourse Y with nine elements. It is determined by the max-algebraic-product composition of E, T, and F. Each element of the output fuzzy set denotes a value for the corresponding control parameters. Figure 14 illustrates the fuzzy neurons defined for Layer 5 and their connections.

4 Simulation Results

Simulation studies have been conducted to demonstrate the performance of the proposed fuzzy neural network for motion estimation. Two main experiments were conducted, each experiment contained different number of simulations. The first consisted of five simulations for moving objects with different speeds and trajectories. The objective of this experiment was merely to test the simulator and evaluate the motion estimator. The second experiment consisted of twenty four simulations for a moving object with six different velocities under four different noise conditions. The objective of this experiment was to measure the error values in estimated average velocities in noise free and noisy environments. The experiments were conducted using a simulation program developed by the author and run on a SPARC station. The fuzzy neural network calculates the 6-bit average velocities in terms of *pixel per frame* (ppf).

We present here two simulations of the first experiment and one simulation of the second experiment. In the first simulation a single object moving around a circle with a constant speed of 0.2 ppf. The background brightness varies continuously from dark to bright. Figure 15 shows the results of the this simulation in which 741 image frames were used. It contains three types of images: input frame (left), output frame(right) and computed average velocities (bottom). In the output frames, the darker the pixel is, the higher is its average velocity. The graph in the bottom section of the figure illustrates the computed average velocities, represented by arrows, at the end of the simulation. The size of each arrow indicates the last average velocity of the corresponding pixel; the smaller the arrow is, the lower is the pixel's average velocity. The direction of the arrow shows the direction of the last movement to the pixel. For pixels with no arrows, the average velocity is zero.

In the second simulation an object travels along a circular trajectory with varying speed. The object starts travelling clockwise from the top of the circle with the lowest speed in this simulation (0.02381 ppf). The object increases its speed continuously. When the object rotates $90°$, the speed reaches 0.2 ppf (which is the highest speed in this simulation). Then the object starts decelerating, and the speed is decreased continuously until the object reaches the bottom of the trajectory. This trend is repeated again until the object returns to the starting point (i.e., top of the circle). A total number of 3624 image frames were used in this simulation. Figure 16 shows the results obtained from this simulation.

The third simulation was conducted on four different noise condition. We employed the gaussian white noise with different signal to noise ratios. There is an object moving horizontally from left to right with the speed of 0.02222 ppf. The simulation results under the following conditions are presented in Figure 17: (a) input images are noise free, (b) input images with 22 dB signal to noise ratio (SNR), (c) input images with SNR = 17 dB , (d) input images with SNR = 12 dB. We present two graphs for each part of the simulation. The graphs indicate percent relative error of estimated average velocities. The left graph shows the percent relative error versus frame number calculated for only the pixels which should receive a motion in each sampling interval; this is to find out the relation between the system performance versus velocity values. The right graph indicates the percent relative error versus frame number averaged for all the pixels in the image; this is to measure the overall performance of the system. We employed 2383 image frames in this simulations.

In the first experiment the obtained results are good and fulfil our expectation. In the second experiment, it can be seen from the relative error graphs that there is a transient error at the beginning of each simulation. This error occurs because the system calculates the average velocity based on previously obtained velocities. Since at the beginning of each simulation, there are no previously calculated velocities, the error is expected to be large. This error disappears quickly as more frames are processed, and decreases more rapidly if the velocity is high. This transient behaviour of the system can be improved by changing the initial values of the variables which store the last velocity values of the pixels. According to the results, for the noise free simulations there is no error in calculation of the velocity vectors (except for transition errors). This shows that the system works well in the absence of noise. The system also works well if the noise level is moderate. As the noise levels increase from moderate to high the system performance deteriorates slowly and gracefully. Based on the simulation results, the fuzzy neural network showed a good performance in detecting moving objects and estimating their velocities. We found out that the performance of the system does not depend on velocity values of the moving objects.

5 Conclusions

A generic model of a fuzzy neuron was defined as an extended model of the existing fuzzy neurons. We mentioned the differences between the generic fuzzy neuron and the other models. Fuzzy relations were employed in the generic fuzzy neuron functions (connection, aggregation, and activation) and the *-* compositional rule of inference was utilized to obtain the solution of the relational assignment equations used in the generic fuzzy neuron. In this way, any type of fuzzy operation in the class of triangular norms or triangular conorms can be utilized to derive the output of a generic fuzzy neuron function. A five-layer feedforward fuzzy neural network was proposed to implement a motion estimation algorithm. We evaluated the performance of the fuzzy neural network in simulation studies. According to the simulation results, the fuzzy neural network showed a significant performance in detection of moving objects and estimation of their velocities.

Appendix A

Definition 1: The *absolute difference* of two fuzzy sets A and B, $A \boxminus B$, is the fuzzy set defined by the following membership function

$$\mu_{A \boxminus B}(x) = |\mu_A(x) - \mu_B(x)| \tag{38}$$

Definition 2: The *algebraic division* of two fuzzy sets A and B, $A \boxslash B$, is the fuzzy set defined by the following membership function

$$\mu_{A \boxslash B}(x) = min\left\{1, \frac{\mu_A(x)}{\mu_B(x)}\right\} \tag{39}$$

Definition 3: If R is a fuzzy relation in $X \times Y$ and S is a fuzzy relation in $Y \times Z$, the *-* *composition* of R and S, $R \overset{*}{\underset{*}{\circ}} S$, is a fuzzy relation in $X \times Z$ defined as follows

$$R \overset{*}{\underset{*}{\circ}} S \leftrightarrow \mu_{R \overset{*}{\underset{*}{\circ}} S}(x, z) = \underset{y}{*}\{\mu_R(x, y) * \mu_S(y, z)\} \tag{40}$$

where * could be any operator in class of triangular norms or triangular conorms.

Appendix B

Nine fuzzy matrices of the nine DIR-FNs in Layer 4.

$$H_1 = \left[\exp\left(-\beta^2 0^2\right) \exp\left(-\beta^2 1^2\right) \exp\left(-\beta^2 3^2\right) \exp\left(-\beta^2 1^2\right) \exp\left(-\beta^2 2^2\right) \exp\left(-\beta^2 4^2\right) \exp\left(-\beta^2 3^2\right) \exp\left(-\beta^2 4^2\right) \exp\left(-\beta^2 5^2\right) \right]$$

$$H_2 = \left[\exp\left(-\beta^2 1^2\right) \exp\left(-\beta^2 0^2\right) \exp\left(-\beta^2 1^2\right) \exp\left(-\beta^2 3^2\right) \exp\left(-\beta^2 2^2\right) \exp\left(-\beta^2 3^2\right) \exp\left(-\beta^2 4^2\right) \exp\left(-\beta^2 5^2\right) \exp\left(-\beta^2 4^2\right) \right]$$

$$H_3 = \left[\exp\left(-\beta^2 3^2\right) \exp\left(-\beta^2 1^2\right) \exp\left(-\beta^2 0^2\right) \exp\left(-\beta^2 4^2\right) \exp\left(-\beta^2 2^2\right) \exp\left(-\beta^2 1^2\right) \exp\left(-\beta^2 5^2\right) \exp\left(-\beta^2 4^2\right) \exp\left(-\beta^2 3^2\right) \right]$$

$$H_4 = \left[\exp\left(-\beta^2 1^2\right) \exp\left(-\beta^2 3^2\right) \exp\left(-\beta^2 4^2\right) \exp\left(-\beta^2 0^2\right) \exp\left(-\beta^2 2^2\right) \exp\left(-\beta^2 5^2\right) \exp\left(-\beta^2 1^2\right) \exp\left(-\beta^2 3^2\right) \exp\left(-\beta^2 4^2\right) \right]$$

$$H_5 = \left[\exp\left(-\beta^2 4^2\right) \exp\left(-\beta^2 4^2\right) \exp\left(-\beta^2 4^2\right) \exp\left(-\beta^2 4^2\right) \exp\left(-\beta^2 0^2\right) \exp\left(-\beta^2 4^2\right) \exp\left(-\beta^2 4^2\right) \exp\left(-\beta^2 4^2\right) \exp\left(-\beta^2 4^2\right) \right]$$

$$H_6 = \left[\exp\left(-\beta^2 4^2\right) \exp\left(-\beta^2 3^2\right) \exp\left(-\beta^2 1^2\right) \exp\left(-\beta^2 5^2\right) \exp\left(-\beta^2 2^2\right) \exp\left(-\beta^2 0^2\right) \exp\left(-\beta^2 4^2\right) \exp\left(-\beta^2 3^2\right) \exp\left(-\beta^2 1^2\right) \right]$$

$$H_7 = \left[\exp\left(-\beta^2 3^2\right) \exp\left(-\beta^2 4^2\right) \exp\left(-\beta^2 5^2\right) \exp\left(-\beta^2 1^2\right) \exp\left(-\beta^2 2^2\right) \exp\left(-\beta^2 4^2\right) \exp\left(-\beta^2 0^2\right) \exp\left(-\beta^2 1^2\right) \exp\left(-\beta^2 3^2\right) \right]$$

$$H_8 = \left[\exp\left(-\beta^2 4^2\right) \exp\left(-\beta^2 5^2\right) \exp\left(-\beta^2 4^2\right) \exp\left(-\beta^2 3^2\right) \exp\left(-\beta^2 2^2\right) \exp\left(-\beta^2 3^2\right) \exp\left(-\beta^2 1^2\right) \exp\left(-\beta^2 0^2\right) \exp\left(-\beta^2 1^2\right) \right]$$

$$H_9 = \left[\exp\left(-\beta^2 5^2\right) \exp\left(-\beta^2 4^2\right) \exp\left(-\beta^2 3^2\right) \exp\left(-\beta^2 4^2\right) \exp\left(-\beta^2 2^2\right) \exp\left(-\beta^2 1^2\right) \exp\left(-\beta^2 3^2\right) \exp\left(-\beta^2 1^2\right) \exp\left(-\beta^2 0^2\right) \right]$$

REFERENCES

[1] M.M. Gupta, G.K. Knopf, "Fuzzy Neural Network Approach to Control Systems," B.M. Ayyub, M.M. Gupta, L.N. Kanal, *Analysis and Management of Uncertainty: Theory and Applications.* Elsevier Science Publishers B.V., 1992.

[2] K. Hirota, W. Pedrycz, "OR/AND Neuron in Modeling Fuzzy Set Connectives," *IEEE Trans. on Fuzzy Systems,* vol. 2, no. 2, pp. 151-161, 1994.

[3] B. Jahne, *Spatio-temporal Image Processing: Theory and Scientific Applications.* Springer Verlag, Berlin New York, 1993.

[4] J.S.R. Jang, "ANFIS:Adaptive-Network-Based Fuzzy Inference Systems," *IEEE Trans. on Systems, Man, and Cybernetics,* vol. SMC-3, pp. 28-44, 1992.

[5] J.M. Keller, H. Tahani, "Implementation of Conjuctive and Disjunctive Fuzzy Logic Rules With Neural Networks," *Int'l J. Approximate Reasoning,* vol. 6, pp. 221-240, 1992.

[6] B. Kosko, *Neural Networks and Fuzzy Systems: A Dynamical Systems Approach to Machine Intelligence.* Prentice Hall, Inc., Englewood Cliffs, NJ, 1992.

[7] A.Z. Kouzani, *A Fuzzy Neural Network and its Application to Motion Detection and Velocity Estimation.* M.Eng.Sc. Thesis, University of Adelaide, Adelaide, Australia, January, 1995.

[8] H. K. Kwan, Y. Cai, "A Fuzzy Neural Network and its Application to Pattern Recognition," *IEEE Trans. on Fuzzy Systems,* vol. 2, no. 3, pp. 185-193, 1994.

[9] S.C. Lee, E. T. Lee, "Fuzzy Neural Networks," *Mathematical Biosciences,* vol. 23, pp. 151-177, 1975.

[10] H. Li, H.S. Yang, "Fast and Reliable Image Enhancement Using Fuzzy Relaxation Technique," *IEEE Trans. on Systems, Man, and Cybernetics,* vol. SMC-19, no. 5, pp. 1276-1281, 1989.

[11] J.N. Lin, S.M. Song, "A Novel Neural Network for the Control of Complex Systems," 1994.

[12] W.S. McCulloch, W.H. Pitts, "A Logical Calculus of Ideas Immanent In Nervous Activity," *Bull. Math. Biophys.,* vol. 5, pp. 115-133, 1943.

[13] A. Mitiche, *Computational Analysis of Visual Motion.* Plenum Press, New York, 1994.

[14] S.K. Pal, R.A. King, "Image Enhancement Using Smoothing with Fuzzy Sets," *IEEE Trans. Syst., Man, Cybern.,* vol. SMC-11, no. 7, pp. 494-501, 1981.

[15] S.K. Pal, A. Rosenfeld, "Image Enhancement and Thresholding by Optimization of Fuzzy Compactness," *Pattern Recognition Letters,* vol. 7, pp. 77-86, 1988.

[16] W. Pedrycz, "Neurocomputing in Relational Systems," *IEEE Trans. on Pattern Analysis and Machine Intelligence,* vol. 13, no. 3, pp. 289-296, 1991.

[17] W. Pedrycz, A.F. Rocha, "Fuzzy-Set Based Modeles of Neurons and Knowledge-Based Networks," *IEEE Trans. on Fuzzy Systems,* vol. 1, no. 4, pp. 254-266, 1993.

[18] H. Takagi, I. Hayashi, "NN-Driven Fuzzy Reasoning." *Int'l J. Approximate Reasoning,* vol. 5, no. 3, pp. 191-212, 1991.

[19] W.B. Thompson, S.T. Barnard, "Lower-Level Estimation and Interpretation of Visual Motion," *Computer,* 14, no. 8, pp. 20-28, 1981.

[20] T. Yamakawa, S. Tomoda, "A Fuzzy Neuron and Its Application to Pattern Recognition," *Proc. Third IFSA Congress,* Washington, pp. 943-948, August 6-11, 1989.

[21] T. Yamakawa, M. Furukawa, "A Design Algorithm of Membership Functions for A Fuzzy Neuron using Example-Based Learning." *IEEE International Conference on Fuzzy Systems, March 8-12, 1992,San Diego, California.* IEEE Press,1992.

[22] L.A. Zadeh, "Fuzzy Sets," *Inform. Control,* vol. 8, pp. 338-353,1965.

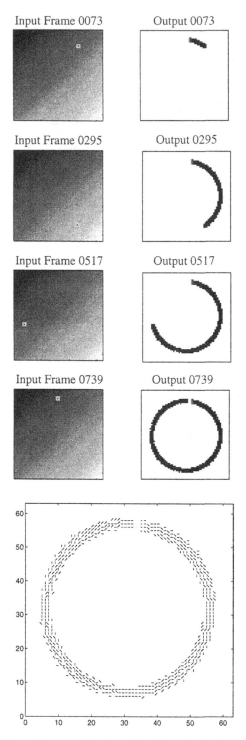

Figure 15: An object moving on a circular path with a constant speed of 0.2 ppf.

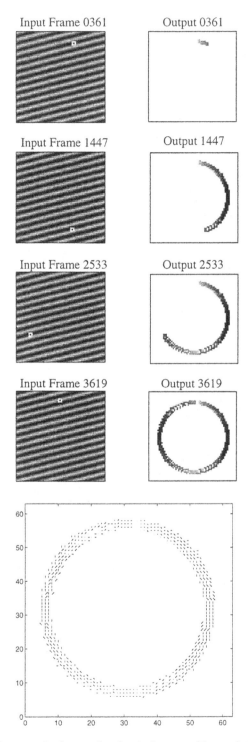

Figure 16: An object travels along a circular trajectory with varying speed.

171

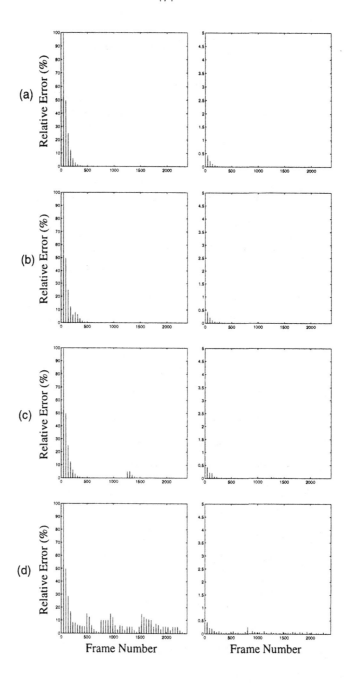

Figure 17: Relative errors for an object speed = 0.02222 ppf. (a) Noise free, (b) *SNR* = 22dB, (c) *SNR* = 17dB, (d) *SNR* = 12dB.

Fizzy-Fuzzy Inferencing

Andrzej Buller
Faculty of Electronics, Technical University of Gdańsk
G.Narutowicza 11/12, 80-952 Gdańsk, POLAND
e-mail: buller@sunrise.pg.gda.pl

The paper presents a method of a fuzzy knowledge processing in a massively parallel way. The key concept is Running Agent (RAT)—a mobile entity carrying a piece of processable information. A Working Memory inhabited by a society of RATs may be implemented in two versions: (i) recurrent neural network in which RATs are represented by navigating patterns, or (ii) a test tube containing a suspension where RATs are molecules of defined structures. The RATs are continuously indoctrinated in such a way, that for a given period of time n_1 RATs are made to have an opinion "A", while n_2 RATs are made to have an opinion "not A", where $n_1/(n_1+n_2)$ equals the assumed degree of membership of a given percept in the fuzzy set A. Some other RATs are made to adhere logical rules related to the A. During a "debate" in the society, concluding opinions are produced. Hence, one may observe a fizzy growth of populations of RATs carrying several contradictory statements. When a poll indicates a domination of one of the populations in the Working Memory, it may be considered as finding of a final solution. The discussed method of inferencing may play in knowledge processing similar role as the Monte Carlo method plays in digital integration. Some results of experiments with a simulation model of a Fizzy-Fuzzy system are also discussed.

1. Introduction

The idea of the Society of Mind introduced by Marvin Minsky [1] strongly influences the recent AI projects intending to design a human-like cognitive systems [2][3][4]. It has been also shown that some types of NP-hard problems, as, for example, the

Hamiltonian Path problem, may be solved in a test tube using DNA strands as data-carriers processed in a random and parallel manner [5]. "Why not to try to design chemicals which would facilitate a symbolic knowledge processing? Why not to try to emulate the content of such a test-tube in such a way that patterns representing molecules would navigate in a neural working memory interacting one with another met?" asks Buller [6]. Massively parallel computation provides the opportunity of employing the new paradigm. Indeed, Kitano argues that "...massively parallel artificial intelligence will add new dimension to our models of thought and to the approaches used in building intelligent systems" [7:2].

Buller [8], recognizing notions' membership in fuzzy sets as a necessary concept, notes that classic fuzzy calculus based on Łukasiewicz-Zadeh set-theoretic [9] or other mathematical operations [10] may be replaced with a competition between populations of mobile agents carrying contradictory statements. A society of the agents is assumed to inhabit a Working Memory which may be implemented as in two versions: (i) recurrent neural network in which the agents are represented by navigating patterns [11][12], or (ii) a test tube containing a suspension where the agents are molecules of defined structures [6]. Each agent—a member of the society —navigates all over the Working Memory and interacts with other such agents. As it was shown in [13][14] and [6], owing to special properties of the agents and the Working Memory, the society is able to work out an appropriate decision based on given percepts and statements stored in a Fuzzy Knowledge Base .

The key concept of the proposed paradigm is Running Agent (RAT)—a mobile entity carrying a piece of processable information. The RATs inhabiting the Working Memory are continuously indoctrinated in such a way, that for a given period of time n_1 RATs are made to have an opinion "A", while n_2 RATs are made to have an opinion "not A", where $n_1/(n_1+n_2)$ equals the degree of membership of a given percept in the fuzzy set A. Some other RATs are made to adhere logical rules related to A. During a "debate" in the society, concluding opinions are produced. Hence, one may observe a fizzy growth of populations RATs carrying several contradictory statements. When a poll indicates a domination of one of the populations in the Working Memory, it may be considered as finding of a final solution. Although Zadeh [15] notes that "it is unnatural to force a voter to choose either A or not A when A is a fuzzy concept", this seems to not concern the presented paradigm which employs artificial voters of very limited intelligence and their specific role in the process of inferencing.

2. Fuzzy Knowledge - Fizzy Inferencing

A Fizzy-Fuzzy system proposed in [6] consists of a Fuzzy Knowledge Base, an Indoctrinator, a Working Memory inhabited by the society of RATs (Running Agents), and a Poll device (see Fig.1). An input/output device for each t belonging to T provides a set of percepts X_1, ..., X_n and receives calculated decisions, where $T = \{1, 2, ...\}$ is a space of discrete values of time.

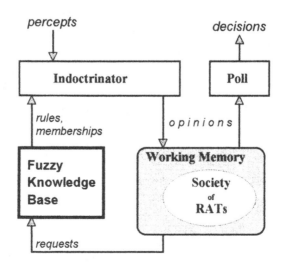

Fig.1. A Society of RATs-based Fizzy-Fuzzy system

RAT is a mobile entity consisting of: (1) a strategy of its own behavior in the Working Memory and (2) its "own" opinion in a matter (Fig.2).

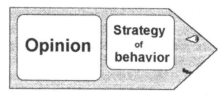

Fig.2. Structure of Running Agent (RAT)

2.1. Fuzzy Knowledge Base

Let us assume that A_1, ..., A_M are facts related to particular percepts, while A_{M+1}, ..., A_{M+m} are to be deduced. All As shall be interpreted as fuzzy sets. Fuzzy Knowledge Base provides the functions Ψ and μ and a set of rules S_1, ..., S_m.

$\Psi(i)$ returns the number of a percept being a member of the fuzzy set numbered i, while $\mu_{A_i}(X_j)$ returns the degree of the membership of a given percept X_j in a given fuzzy set A_i. S_i is a pair $\langle y_i, Y_i \rangle$, where y_i belongs to the set $\{A_{M+1}, ..., A_N\}$, while Y_i is a subset of $\{A_1, ..., A_N\}$. An ultimate version of the Knowledge Base will provide only "rules on request", which would keep unrelated statements purposefully unavailable.

2.2. Indoctrinator

Indoctrinator generates statements to be carried by RATs as their opinions. Its output is described by the function:

$$g|T \rightarrow \emptyset \cup \{A_1, ..., A_n\} \cup \{\neg A_1, ..., \neg A_n\} \cup \{S_1, ..., S_m\}$$

such that for j=1, 2, ..., K:

$P(\,g_j(t) \neq \varnothing\,) = r,$

$P(\,g_j(t) = f_i(t)|\,g_j(t) \neq \varnothing\,) = e_i/(e_1 + e_2 + ... + e_{M+n}),$

where:

K - number of inputs to the Working Memory,

P - probability,

r - density of stream of opinions,

$e_1 + e_2 + ...$ - importance factors,

$f_i|T \to \{A_i, \neg A_i\}$ - an auxiliary function (see Fig.3), such that for i=1, 2, ..., M:

$P(\,f_i(t) = A_i\,) = \mu_{A_i}(X_{\Psi(i)})$

$P(\,f_i(t) = \neg A_i\,) = 1 - \mu_{A_i}(X_{\Psi(i)})$

while for i=M+1, M+2, ..., M+m,

$f_i(t) = S_{i-M} = \text{const.}$

The parameters r, as well as e_1, e_2... are to be set up by user. However, as it was suggested in [6], a knowledge-based self-tuning may be also considered.

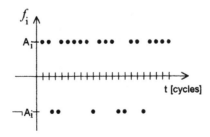

Fig.3. The exemplary plot of the discrete function f_i. (The probability that for a given moment t the function returns A_i equals the degree of the membership of a percept in the fuzzy set A_i.)

2.3. Working Memory

Working Memory constitutes an environment in which the society of RATs lives. Let us consider a model of Working Memory being an array of processing nodes, where each node can receive/send binary words from/to its nearest neighbors (or from/to itself, especially when a given neighbor is damaged or does not exist at all). According to a form and sophistication of the information to be processed, the nodes may be implemented either as fully designed digital systems or as neural networks. In this paper a 2-dimensional 4-connected environment (i.e. in which each processing node has bi-directional connections to four neighbors) is discussed (Fig.4). Of course there is nothing magic in the number "4". Moreover, one may suppose that machines, say, 6- or 8-connected or three-dimensional 6-, 8-, or 12-connected would work much better. On the other hand, one may note that all other structures would require decidedly more sophisticated strategies of RAT behavior.

Generally, every node is to (i) admit a RAT incoming at a given time and send it to one of attainable neighbor-nodes according to the strategy the RAT is endowed with, or, when two or more RATs enter the node at the same time, (ii) facilitate appropriate RAT-RAT interactions. Regardless the number of inter-node connections, RATs, jumping from one node to another, navigate all over the Working Memory.

The Indoctrinator provides RATs with opinions. Some of the opinions may be contradictory. A given RAT may navigate passing consecutive nodes and interact with other RATs it can meet. Especially essential is the interaction between a RAT carrying a rule and a RAT carrying a related fact. In such case the first RAT will replace the rule with a logical conclusion from the opinions of the two RATs. Such interactions may take place in several nodes at the same time. Hence, in a period of time one may observe an appearance of a group of RATs having an opinion which might be recognized as a final conclusion, as well as RATs having an opposite opinion.

Since new, deduced opinions appear continuously in the Working Memory in a large number of their copies, it resembles the appearance of bubbles during a turbulent chemical reaction. Hence, Phil Davies called such process a "fizzy" calculus [16]. After a period of unstable situation, adherents of a particular solution should decidedly win, unlike the cases in which a satisfactory decision does not exist at all.

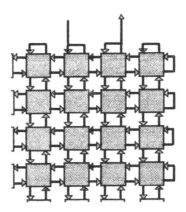

Fig.4. An exemplary structure of Working Memory. RATs, passing consecutive nodes, navigate all over the mesh and interact one with another met.

2.4. Poll

The Poll device is to collect continuously RATs' opinions, and, based on a mean results for a defined period of time, to work out appropriate decisions.

3. Strategy of RAT behavior

The first efficient strategy of agent's behavior was discovered via gradual improving of an inefficient initial strategy by a process of "trial and error" using a simulation model of Jigsaw Machine [13]. The strategy has been formalized for further analysis [12][14]. It consists of the two rules: Navigation Rule and Interaction Rule described here in the form of "guidelines" addressed to each RAT living in a 2-dimensional 4-connected Working Memory [cf. 6].

3.1. Navigation Rule (*N*-Rule)

Assuming that:

- having entered a processing node you can see a left-hand gate, a gate in front of you, a right-hand gate and a gate behind you (i.e. the gate you entered through),

- E is a set of RATs you have met in the processing node, while R_L, R_F, and R_R are the RATs who might enter the node through your left-hand gate, the gate in front of you, and your right-hand gate respectively,

- Q is a 3-bit binary word "$q_1q_2q_3$" such that: $q_1=1$ iff $R_L \in E$, $q_2=1$ iff $R_F \in E$, and $q_3=1$ iff $R_R \in E$, and

- F denotes the action {$\alpha \leftarrow \varnothing$, then apply *I*-Rule, and then **go forward** (i.e. exit the node through the gate in front of you)}, L and R denote the action {$\alpha \leftarrow R_L$ or $\alpha \leftarrow R_R$ respectively, then apply the I-Rule, and then **turn right** or **turn left** respectively (i.e. exit the node through your right-hand gate or left-hand gate, respectively)}, and B denotes the action {$\alpha \leftarrow R_F$, then apply the *I*-Rule, and then **turn back** (i.e. exit the node through the gate behind you)},

follow the guidelines:

> if Q=101 or Q=000 then F;
>
> if q_2 then B;
>
> if Q=100 or Q=001 then R or L respectively.

If we also assume that the value of Q denotes (in the classic binary code) a position in a string, the N-Rule will take on a shape quite convenient for a genetic processing, i.e.

FLBBRFBB

The N-Rule is not necessary in chemical Working Memory but in case of neural Working Memory it facilitates changes of the direction of RAT motion as often as an appropriate condition is satisfied. This resembles Brownian movements. Hence, the RAT has a good chance to meet an appropriate mate in a reasonable time.

3.2. Interaction Rule (I-Rule)

Assuming that

- an integer G (G-factor) initially equal to 0 is assigned to each RAT at the moment of its joining the Society,

- α is the RAT of reference defined in the N-Rule,

- y and Y are "your y" and "your Y" respectively,
  while y' and Y' are "α's y" and "α's Y" respectively,

- Q' is a 4-bit binary word $q_1 q_2 q_3 q_4 q$ such that:
    $q_1=1$ iff $y' \notin Y$ and $\neg y' \notin Y$,
    $q_2=1$ iff $y \notin Y'$ and $\neg y \notin Y'$ and (G>G' or Y=Y'=$\varnothing$ and y=$\neg y'$),
    $q_3=1$ iff Y'=$\varnothing$, and
    $q_4=1$ iff $y' \in Y$,

- F denotes the action "**kill** α", L denotes the action $\{G \leftarrow G+1, \text{ then } Y \leftarrow Y\text{-}y'\}$, and R denotes the action $\{G \leftarrow G+1, \text{ then } y \leftarrow \neg y, \text{ then } Y \leftarrow \varnothing\}$,

follow the guidelines:

> if q_1 then { if q_2 then F
> else if q_3 then { if q_4 then L else R }.

If we also assume that B denotes the action "**do nothing**", while Q' denotes (in the classic binary code) a position in a string, the *I*-Rule will take on the shape:

BBLRBBLRBBBBFFFF

A full DNA-like code determining RAT behavior may therefore take on the shape:

FLBBRFBBBBLRBBLRBBBBFFFF

where 8 first "bases" constitutes the "gene" responsible for RAT navigation, while the rest of the strand constitutes the "gene" determining interactions with other RATs.

The *I*-Rule describes both the behavior of enzymes which should be designed to facilitate interactions between RATs represented by the true DNA strands and the requirements for processing nodes of a neural Working Memory.

To prevent the society of RATs from a congestion, non-prospective RATs are eliminated based on the G-factor values. The higher G, the bigger chance for survival, as an RAT of higher G can kill the encountered one when their informational contents have nothing in common. G increases after each successful production of a concluding information.

4. Example

Let us assume that the Fuzzy Knowledge Base contains, the rules:

"alarm if not ok",
"ok if warm and not hot",

which, formally, may be denoted as:

$S_1 = \langle alarm, \{\neg ok\}\rangle$
$S_2 = \langle ok, \{warm, \neg hot\}\rangle$.

The used notions:

A_1 = "warm", $\quad A_2$ = "hot",
A_3 = "ok", $\qquad A_4$ = "alarm"

are considered as fuzzy sets.

A_1 and A_2 are defined in such a way that, say:

$$\mu_{warm}(45°C)=.90, \qquad \mu_{hot}(45°C)=.10,$$
$$\mu_{warm}(50°C)=.95, \qquad \mu_{hot}(50°C)=.30,$$
$$\mu_{warm}(53°C)=.94, \qquad \mu_{hot}(53°C)=.41,$$
$$\mu_{warm}(55°C)=.90, \qquad \mu_{hot}(55°C)=.50,$$

Since the input/output device provides the one-element set of percepts and a given percept may be a member of the two defined fuzzy sets, $\Psi(1)=\Psi(2)=1$.

Given a particular measured temperature the question is:

"To trigger alarm or not?"

For r=0.5 there is a probability 0.5 that at a given moment the Indoctrinator provides an opinion to a given communication channel in the Working Memory. The probability that it will be a particular opinion depends on the importance factors. For the importance factors $e_1=2$, $e_2=2$, $e_3=1$, $e_4=3$ and the given temperature T=55°C, the distribution of the probability is shown in the table:

opinion	probability
$\langle$warm$\rangle$	0.225
$\langle\neg$warm$\rangle$	0.025
$\langle$hot$\rangle$	0.125
$\langle\neg$hot$\rangle$	0.125
$\langle$alarm, $\{\neg$ok$\}\rangle$	0.125
$\langle$ok, $\{$warm,$\neg$hot$\}\rangle$	0.375

If an RAT comes to the communication channel at the same time as the opinion from the Indoctrinator, the opinion it carried will be replaced with the newly provided one. If at the moment the channel is empty, a new RAT will be generated to carry the newly provided opinion. All RATs navigate and exchange their opinions according to the strategy of behavior each RAT has in its DNA-like code. After a period of time a poll concerning the RATs having the opinion $\langle$alarm$\rangle$ or $\langle\neg$alarm$\rangle$ is conducted. The result of the poll shall answer the question "To trigger alarm or not?"

5. Experimental results

The above example has been tested using a simulation model of the Working Memory and the Society of RATs written in Split-C and run on the CM-5 Connection Machine.

The results for a 10×10-node 4-connected Working Memory and the strategy:

```
FLBBRFBBBBLRBBLRBBBBFFFF
```

has been described in [14]. Note that the strategy does not allow the situation that two or more RATs want enter the same channel at the same time. On the other hand, the strategy prevents RATs from visiting certain areas in the Working Memory. Despite this the results seemed to be quite promising. The further experiments with 25×25-node Working Memory led to a similar conclusion.

In the recent experiment described in [6] a 25×25-node Working Memory with four opinion inlets has been simulated. All data was as in the example described in the section 4. A bit different strategy of RAT behavior has been employed. In the way of an arbitrary mutation the 3rd base B was replaced with L, which gave the strand:

```
FLLBRFBBBBLRBBLRBBBBFFFF
```

Owing to this every RAT had a chance to visit any node of the Working Memory.

Fig. 5. shows the varying poll results for 5 consecutive experiments conducted for each of the 4 different percept values. n_{alarm} is percentage of RATs having opinion "**alarm**" in the population of RATs having opinion "**alarm**" or "**not alarm**". The vertical lines marked N_{alarm} shows the membership of the given T to the fuzzy set alarm, calculated using classic set-theoretic operations and multiplied by 100.

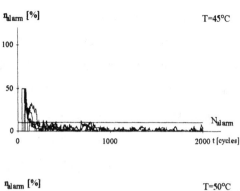

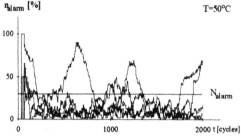

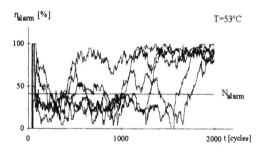

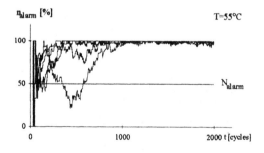

Fig.5. Poll results varying in time (five independent experiments for each of the presented temperatures). The vertical lines marked N_{alarm} show in each of the charts the degrees of membership of given T in the fuzzy set "alarm" obtained using classic set-theoretic operations and multiplied by 100.

One can see that for T=45°C and T=55°C, after a short period of unstable situation, the Fizzy-Fuzzy Calculus is more decisive than the degree of membership in the fuzzy set "alarm" calculated using the classic set-theoretic operations. For T=50°C and T=53°C the poll results are quite random. However, in these cases the classic set-theoretic operations also provides no basis for a decision. Nevertheless, based on the Fig.5., it may be noted, that, if for a given temperature an average value of the 5 independent poll results were taken, the decisiveness of the fizzy-fuzzy calculus in case of T=50°C and T=53°C would be significantly improved.

6. Concluding remarks

In the age of cheap chips, one may anticipate the possibility of employment of massively parallel algorithms on the level of small single controller. The proposed paradigm replaces classic fuzzy calculus with a competition between populations of RATs carrying contradictory statements and looks, based on simulation experiments, more decisive than the classic fuzzy calculus. It may play in knowledge processing similar role as the Monte Carlo method plays in digital integration. An implementation of a fizzy-fuzzy inferencing system may be considered as either a recurrent neural network or a chemical reactor.

References

1. Minsky M (1986) The Society of Mind, Simon and Schuster.

2. Brooks R A, Stein L A (1994) "Building brains for bodies", Autonomous Robots, 1 (1), 7-25.

3. Cheeseman P (1995) "Yet another cognitive architecture—modules, models, money and markets", a talk at the Symposium: Representing Mental States and Mechanisms (AAAI Spring Symposium Series, Stanford University, March 27-29, 1995).

4. Buller A. (1995) "Operations on Multimodal Records: Towards a Computational Cognitive Linguistics", Technical Report TR-95-027, International Computer Science Institute, Berkeley.

5. Adleman L. M. (1994) "Molecular computation of solutions to combinatorial problems", Science, 266 (5187) 1021-3.

6. Buller A. (1995) "Fuzzy vs. Fizzy", Proceedings of the 1955 IEEE/Nagoya University WWW on Fuzzy Logic and Neural Networks/Evolutionary Computation, Nagoya, Japan, 1-6.

7. Kitano H. (1994) "The Challenge of Massive Parallelism", In: H. Kitano and J. A. Hendler, Massively Parallel Artificial Intelligence, AAAI/MIT Press, 1-51.

8. Buller A. (1992) "Fuzzy Inferencing as a Competition Between Contradictory Statements", The Proceedings of First Singapore International Conference on Intelligent Systems, 103-106.

9. Zadeh L. (1965) "Fuzzy Sets", Information and Control, 8, 338-353.

10. Zimmerman H. -J. (1991) Fuzzy Sets Theory - and Its Applications, Kluwer Academic Publishers.

11. Buller A. (1990) "A Sub-neural Network for a Highly Distributed Processing", In: C. N. Manikopoulos (ed.), Proceedings of 8th International Congress of Cybernetics and Systems}, New York, June 11-15, 1990, (New Jersey Institute of Technology Press), Vol.1, 185-192.

12. Buller A. (1993) "Neural Screen and Intelligent Patterns", Proceedings of 1993 International Joint Conference on Neural Networks (IJCNN'93-NAGOYA), 379-382.

13. Buller A., Davies P. (1991) "On a Jigsaw Architecture for Logic Programming and Pattern Recognition", Technical Report CS-34-91, Department of Computing, University of Bradford.

14. Buller A. (1995) "Fuzzy Inferencing: A Novel Massively Parallel Approach", Technical Report TR-95-025, International Computer Science Institute, Berkeley.

15. Zadeh L. (1995) "Discussion: Probability Theory and Fuzzy Logic are Complementary Rather Than Competitive", Technometrics, 37 (3), 1-6.

16. Davies P. (1991) - personal communication.

Multi-layer Perceptron Design Using Delaunay Triangulations

Elena Pérez-Miñana, Peter Ross, John Hallam

Department of Artificial Intelligence, University of Edinburgh, 80 Southbridge, Edinburgh EH1 1HN, Scotland, U.K.

Abstract. The successful development of an application using the multi-layer perceptron (MLP) model greatly depends on the structural complexity of the domains involved. Different mathematical and/or statistical techniques can be used to subtract the maximum amount of information of this type from an available sample of the input space. In the context of the MLP model, it has been used to decide on the form the parameters of the network and/or related learning algorithm (LA) should have. This paper describes the information subsumed in the Delaunay triangulation (DT) and Voronoi diagram (VD) of the points comprising the input space of an application, how it might be used to evaluate the convenience of building a network based on the MLP model for its implementation and to estimate an initial architecture that can be subsequently improved by a pruning process.

1 Introduction

One of the main aims of an application developer is that network performance not be restricted to the training set (Ts). The network should correctly generalise to additional patterns. One way of achieving this is to design the smallest architecture that will fit the data. There are many methods for designing a MLP (see [7], [15] for reviews) but none provides a guarantee of a smallest architecture. Indeed the generation of optimal networks is **NP-complete** [8].

MLPs are generally agreed to be specially useful for the automatisation of classification tasks. This is partly because this type of task can easily be expressed as a mapping from groups of input values to values in a discrete and well-specified range. Also, Cybenko has shown that a one hidden layer MLP can approximate any continuous function if enough units are provided for the hidden layer [4].

Cybenko's key condition, that "enough hidden units" be employed, can also be expressed as there being enough surfaces to correctly partition the input space. One way of complying with this is by examining its structure, an important step in many current dynamic network configuration algorithms [1, 2, 6, 9, 13, 14]. A certain group of geometric constructs which have recently become important concepts in the area of computational geometry, the Voronoi Diagram (VD) and its dual the Delaunay Triangulation (DT) [12] have proven particularly effective in this context as they both provide information on the degree of proximity of the points in an input space, making it possible to assess the degree of difficulty involved in generating a suitable network. The same information can also be used to compute "reasonable" estimates of a maximal initial architecture. As the output of this analysis is "maximal" a second stage must be incorporated in the design process, one providing information on "superfluous" units and/or connections which might be pruned. A convenient technique to perform this task is principal component analysis because, as it was specifically designed to detect the most important dimensions of an input space relative to a set of points, it is an effective manner of eliminating hidden units, if each is considered to be a dimension of the space provided as input for the data analysis under consideration.

In addition to applying the information subtracted from the VD and DT of a set of points for constructing a maximal network architecture, it is also possible to employ it for measuring the appropriateness of the model for building the application of interest. It is feasible to do this given that, a reliable estimate of the degree of complexity associated to the structure of the domain under scrutiny, can be computed from these geometrical constructs. Taking into account, that the term "complexity" in this study, pertains to the level of difficulty associated to the task of identifying the class to which a particular point belongs, it is reasonable to affirm that knowledge about the spatial disposition of the points comprising a sample of the domain of interest provides an effective tool for quantifying such an attribute.

Finally the relevance of computing the "complexity" measure described should be noted. As it is usually the case that a problem that wants to be solved through an application will require modeling the domains involved and the mapping that exists between them and, the vast amount of models that are available to perform this operation, the possibility of evaluating whether a particular model is appropriate for a specific case, is a very useful tool for any development process.

The techniques described in the preceding paragraphs can be organised into the set of steps comprising a complete design methodology for developing applications using the MLP model. As a detailed discussion of such method is beyond the scope of this publication, the following sections are only concerned with an explanation of how the information provided in the VD and DT is exploited in the network design. Section II describes the type of information available in both geometrical constructs, how they relate to one another and how they can be employed in designing MLP architectures. Section III includes some experiments showing the usefulness of this new type of analysis. Section IV comprises the conclusions reached with this study.

2 Delaunay Triangulations & Voronoi Diagrams

VD and DT are a type of tessellation [3] whose main distinguishing property is that both are totally derived from the data. The VD generated from a set of M points, $\{P_1, P_2, ..., P_k, ..., P_M\}$ in N-dimensional space, called sites, corresponds to a partition of the space into convex regions, called Voronoi cells. Each cell corresponds to a "region of influence" surrounding each site. In two dimensions the Voronoi cells are convex polygons whose edges correspond to the perpendicular bisectors of the Delaunay triangle edges joining neighbouring sites. The triangulation results from connecting each pair of data sites that share a common Voronoi region boundary. Extending the concept to higher dimensions a Voronoi cell is an N-dimensional polytope which can be defined as the intersection of a finite number of closed half-spaces and is, therefore, supported by a finite number of hyperplanes. For a set of N-dimensional points its dual, the DT, can be obtained from the projection of the "lower" hull of these same points transformed into an (N+1)-dimensional space.

The DT and VD of a Ts provide a very detailed description of the degree of existing proximity of the points included therein given the conditions each must satisfy. An extensive analysis of the properties and applications of these geometric constructs is not the main objective of this publication and can be found in [10], therefore an outline of the general definition of the main underlying concept should be sufficient, at least for obtaining a clearer understanding of the manner in which the theory involved is applied in the development of a network based on the MLP model. As was previously mentioned the VD is comprised of a number of "voronoi regions" ($VR_i, i \in [1, M]$) one for each of the elements (sites) in the set from which it is being derived . The voronoi region VR_i associated to the site P_i is given by:

$$VR_i = \{P/||P - P_i|| <= ||P - P_j||\}$$
$$j \neq i; i, j \in [1, M]$$

The term $||P - P_i||$ corresponds to the euclidean distance that exists between the points P and P_i. The above means that each VR_i is conformed by all those points in the N-dimensional space whose closest site of the input set is P_i. The boundaries of a region VR_i correspond to ridges, eg. edges in the plane, and include those points which are equally close to the sites sharing the ridge. Knowing which sites share boundaries allows to infer information concerning their proximity which can be successfully exploited in the network design process. Given that the DT is the dual of the VD, it provides the same information but because it is organised in a different manner it can be employed in a very particular way in the context of network development using the MLP.

Bose and Garga in [2] describe a method for designing a MLP for classification tasks that combines the information contained in the VD of the Ts with certain concepts subtracted from graph theory. This integration resulted in a set of rules that provide very specific means for deciding on the number of hidden units, layers and on the values of the weights of a network architecture. The relevant concepts underlying their method can be summarised as follows:

1. A layer of N hidden units can implement N distinct hyperplanes. Given that any VR_i can be expressed as the intersection of a finite number of closed half-spaces, each of which can be associated with a hyperplane, it is possible to reproduce VR_i with a MLP network having two layers of units with one

unit in the upper layer. It is possible to compute the intersection of the hyperplanes using the logical AND operator.

2. The VD of the Ts is comprised of various regions, one for each site, some of which might be overlapping. This means it is possible to reproduce the distinct ones with a MLP network similar to the one described in **1** but, there will be as many units in the upper layer as there are distinct regions in the VD under consideration.

3. Given that the network implemented is to be associated to a classification task, an extra layer of units must be defined. It will contain as many units as the number of classes involved in the classification process. The appropriate class associated to each of the VR_i is computed by the union of the relevant regions which through the implementation of the logical OR operator will activate the correct output unit.

The previous set of considerations clearly induces an adequate architecture that requires no further training or adjustment because each of the units and/or weight vectors that is incorporated into it is precisely asserted. The main problem with the process is the complexity of the resulting architecture. It is geometrically justified but because it models so minutely the structure subsumed in the VD of the Ts, it probably results in a very poor generaliser, something which explains the fact that this network property is hardly mentioned in [2]. Also, there is no need to implement all the distinctive VR_i, given that only those associated with sites belonging to different classes and which are hard to differentiate due to their proximity are the really problematic cases for the MLP network. This means that the theory underlying Bose's method will generally produce an architecture with a certain amount of redundant information for which certain measures should have been taken and it is something not accounted for in [2].

Further applications of concepts from computational geometry can be found in [6], [11]. In general they are not concerned with MLPs and employ geometric information only to decide on the type of network architecture.

As the edges of the DT connect the different sites, it is easier to appreciate the degree of intersection that exists between the points of the different classes comprising the Ts. This permits to assess the convenience of using the MLP as a knowledge representation tool and helps in estimating the number of layers that the network architecture should have. The VD provides information for computing the initial number of hidden units required and the values of the weight vectors associated to the input-hidden layer connections. These can be computed by selecting those edges of the VD that separate points belonging to different classes and calculating the direction cosines of the lines associated to each of these edges. The initial set of weight vectors obtained might be reduced by clustering those which are sufficiently aligned and taking one of them as cluster exemplar. The resulting architecture can be considered "maximal" in the sense that for the set of points comprising the Ts the LA will not require too many iterations to achieve complete recognition. It will be using a number of hidden units directly proportional to the number of "hard" cases and at least one layer of connections, i.e. input-hidden, will be initialised in accordance to the

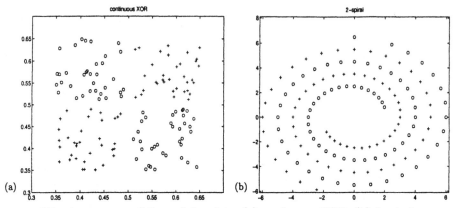

Fig. 1. structure of the training data: (a) continuous XOR, (b) 2-spiral

disposition of the points in the input space. The connections of the other layers can be set to random values as is the usual policy in many network developing processes, in this manner a more flexible architecture is obtained, at least one which will be more effective for correctly classifying patterns not included in the Ts employed for computing the network architecture.

To reduce the redundancy which results from incorporating as many hidden units as there are conflicting regions in the DT of the points in the Ts, given the difficulty of precisely defining the term "conflicting region", it was necessary to incorporate a pruning step in the complete design process, which is the most effective way of dealing with "superfluous" units, something Bose and Garga should have considered. Finally to fine tune the weights which have been randomly initialized a conventional LA, quickpropagation [5] is applied to the network architecture, and it achieves convergence in a reasonable number of iterations, at least while there are enough hidden units in the network, which is the criterion employed in the design process for deciding to stop the pruning stage of the architecture. Quickpropagation was selected for the specific network training because it operates on a fixed architecture but adjusts the LA's parameters according to the way in which the particular execution is progressing and many studies have shown, [7, 15, 5], how important it is to adjust these parameters to the problem being tackled. The examples in the next section explicitly show the advantages of pursuing the steps described and, exploiting the information subtracted from the geometrical constructs of interest.

3 Experimental Results

There are many possible ways of measuring the degree of intersection that exists between the points of the different classes that comprise the input space of interest. For this study the ones considered were: a) generating the DT of the points belonging to each class and subsequently computing how much intersection there is between the structures, i.e. their facets; b) generating the DT of the complete set of points without differentiating between classes and afterwards calculating how many of the edges contain end points belonging to different classes. The experiments show how this information can help a designer decide whether a MLP is an appropriate representation and, for those cases in which it is adequate, how many hidden layers should comprise the network architecture.

The problems selected were the 2-spiral and the continuous XOR. They were chosen because they constitute important benchmarks for evaluating both the representational scope of a network model and the performance level of a LA. Figure 1 shows the training data for both problems. Both constitute hard problems, particularly the 2-spiral case which as can be seen will require many surfaces to correctly partition the input space. This means a MLP architecture with a complexity directly proportional to the size of the Ts thereby increasing the possibilities of producing a network overfitting the training data, i.e. a bad "generaliser". This situation makes the MLP model not well suited for solving the 2-spiral problem even though for the particular Ts employed to build the network architecture it was possible to achieve complete recognition of the training set.

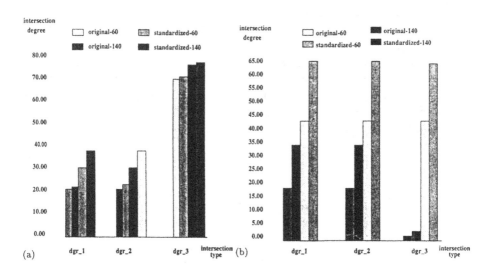

Fig. 2. Geometric information of different sets considered: (a) continuous XOR, (b) 2-spiral

Standardising the set of points of the input space does not vary their relative disposition but often eases the application of a MLP employing a sigmoid activation function. Such transformation usually improves the performance of quickpropagation during subsequent training. However, in the 2-spiral case, the training process was not improved at all even after transforming the data. Therefore to show the effectiveness of the complexity measure derived from the DT the same computation was performed on both original and standardised data. Figure 2 shows the values obtained for both problems with sets of different sizes (60 and 140) so as to determine the influence of Ts size on our analysis. The computation of the measures plotted therein required counting the objects comprising the convex hulls of the different DT generated. For the examples described, given that both problems involve the classification of points into one of two classes, three DT had to be computed, one for the points of each of the classes and a "general" DT obtained from the complete set of points. The quantities of relevance for this computation were:

$$f_i = \# \ facets \ of \ class \ i$$
$$e_{ij} = \# \ edges \ connecting \ class \ i \ to \ class \ j$$
$$f_{ij} = \# \ facets \ of \ class \ i \ intersecting \ those \ of \ class \ j$$
$$i, j \in (1, 2),$$

the edges e_{ij} are in the DT of the complete set of points.

In figure 2 the x-axis is associated with the type of measure being plotted labeled as dgr_1, dgr_2 and dgr_3. Each accounts for some form of intersection: a) dgr_1 is the average of the degree of intersection associated with each of the facets comprising the convex hull computed from the elements of class 1, i.e. with what percentage of the set of facets associated with the points of class 2 did it have a non-null intersection?; b) dgr_2: similar to type 1 but the measure is for each of the facets in the convex hull of the other class; c) dgr_3: the percentage of edges which connect points belonging to only one class for the convex hull obtained from the complete set of points without distinguishing between classes. The equations for the above terms have the following form:

$$dgr_1 = \frac{f_{12}}{f_1}$$

$$dgr_2 = \frac{f_{21}}{f_2}$$

$$dgr_3 = \frac{(e_{11} + e_{22})}{(e_{11} + e_{22} + e_{12})}$$

The best bars in figure 2 are associated with the continuous XOR problem. It shows the lowest dgr_1, dgr_2 and the highest dgr_3, the whole plot is a clear indicator of an appropriate input space for a MLP. By contrast the bars associated with the 2-spiral problem show the complete opposite, i.e. high values associated to dgr_1, dgr_2 and very low values for dgr_3. The size of the problem does not greatly affect the relation between the different types of value computed therefore they can be considered adequate from this point of view.

The fact that the standardised version of the data will be easier to train in the case of the continuous XOR is reflected in two things: a) lower degrees of

Table 1. Performance of trained networks on test sets

Problem	Architecture	%Test set	MSE
continuous XOR	2-7-1	100	7×10^{-3}
2-spiral	2-22-7-1	42.50	3×10^{-1}

intersection in the standardised data, b) greater difference between the values associated with each class and the average computed from the clean edges of the "general" DT, it is this piece of information the one that allows to conclude the degree of complexity of the required network, a smaller value means a larger architecture, below a certain level more hidden layers and a negative difference, as is the case for the 2-spiral problem, indicates that the MLP is not a good representation form.

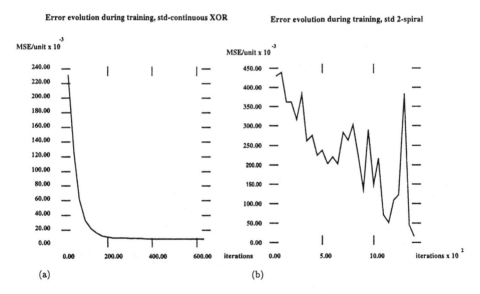

Fig. 3. error evolution during training, x-axis iterations, y-axis MSE/unit: (a) continuous XOR, (b) 2-spiral

The error curves plotted in figure 3 show that the network designed for the standardised continuous XOR data is much easier to train than the one for the standardised 2-spiral problem. Table 1 summarises network performance on test sets for both cases. It clearly shows that with the method described it is possible to train networks to recognise the complete set but for the 2-spiral problem even though the architecture subtracted from the geometric analysis was able to completely and correctly identify the patterns of the Ts, given that the performance achieved on the test set was so poor,that the architecture required is much more complex and most important, the contents of the bar chart describing the "degree of complexity" of the operating domain shown in 2 it is possible to conclude that for this particular example there should be more convenient representation mechanisms than those provided by the MLP model.

4 Conclusions

Our experiments clearly show the value of analysing the structure of the input data prior to network development. Using the knowledge contained in the DT computed from a Ts one can, in the first instance, estimate the degree of complexity of the structure. If the evaluation results are positive it can be used in conjunction with the VD, computed from the same set, to build an initial maximum architecture. By applying an appropriate pruning method one can then achieve a reasonable network for the task of interest. Although our experiments were restricted to a 2-dimensional space, the geometric constructs we employed can be extended to higher dimensions. Our next research objective is to evaluate our methodology on such larger problems giving particular attention to the computational costs the whole process involves which, given the efficiency of the algorithms employed for generating the DT and VD should not be too high when compared against the advantages gained through their use in the network design.

References

1. Alpaydin, E.: GAL: Networks that grow when they learn and shrink when they forget. Technical Report TR-91-032: International Computer Science Institute, (1991).
2. Bose, N.K., Garga, A.K.: Neural network design using Voronoi diagrams. IEEE Transactions on Neural Networks, **4, 5,** (1993) 778–787
3. Bowyer, A., Woodwark, J.: Introduction to computing with geometry. Winchester: Information Geometers, (1993)
4. Cybenko, G.: Approximation by superpositions of a sigmoidal function. Mathematics of Control, Signals, and Systems, **2,** (1989), 303–314
5. Fahlman, S.: An empirical study of learning speed in backpropagation networks. Technical Report CMU-CS-88-162, Carnegie Mellon University, (1988)
6. Fritzke, B.: Growing cell structures - a self-organising network for unsupervised and supervised learning. Technical Report TR-93-026, International Computer Science Institute, (1993)
7. Hertz, J., Krogh, A., Palmer, R.: Introduction to the theory of neural computation. Addison-Wesley Pub., (1991)
8. Judd, P.: Neural networks and the complexity of learning: Cambridge MA: MIT Press, (1990)

9. Karnin, E.D.: A simple procedure for pruning back-propagation trained neural networks. IEEE Transactions on Neural Networks, **1**, **2**, (1990) 239–242

10. Okabe, A.,Boots, B., Sugihara, K.: Spatial tessellations: concepts and applications of Voronoi diagrams. Wiley series in Probability and Statistics, John Wiley & Sons, (1992)

11. Omohundro, S.M.: Geometric learning algorithms. Technical Report TR-89-041, International Computer Science Institute, (1989)

12. O'Rourke, J.: Computational geometry in C. Cambridge University Press, (1994)

13. Reidmiller, M.: Advanced supervised learning in multi-layer perceptrons, from backpropagation to adaptive algorithms. Int. J. of Computer Standards and Interfaces, Special Issue on Neural Networks, **5**, (1994)

14. Romaniuk, S.G., Hall, L.O.: Divide and conquer neural networks. Neural Networks, **6**, (1993), 1105–1116

15. Schiffmann, W., Joost, M., Wierner, R.: Optimisation of the backpropagation algorithm for training multilayer perceptrons. Neuroprose ftp site, (1992)

A Genetic Algorithm for Planning Coal Purchase of a Real Electric Power Plant

Masahiro Inuiguchi[1], Tadahiro Miyake[1],
Masatoshi Sakawa[1], Isao Shiromaru[2]

[1] Hiroshima University, 4-1, Kagamiyama 1-chome, Higashi-Hiroshima, 739, Japan
[2] The Chugoku Electric Power Co. Inc., 4-33, Komachi, Naka-ku, Hiroshima 730-91, Japan

Abstract. In this paper, we focus on coal purchase planning in a real electric power plant. Several complex constraints as well as multiple objectives are involved in the planning problem. The conventional integer programming approach is not suitable for this problem because of its complexity. The versatility of genetic algorithms is exemplified through solving the planning problem. Compared to a simple random search and a simulated annealing, the advantages of a genetic algorithm are shown by numerical simulations.

1 Introduction

The fuel inventory control is an important task in an electric power plant. The purchase order of the fuel is basically made according to an annual sales contract. Thus, a good annual purchase planning can lead to an efficient fuel inventory control. In many real purchase planning problems, there are several complex conditions and criteria. Those complex conditions sometimes make the conventional integer programming technique incompetent because the formulated integer programming problem is too large and difficult to solve in a practical amount of time. Moreover, such a real world problem includes several domain-specific requirements that render the conventional approach not applicable.

In this paper, we treat such an annual coal purchase planning problem for a real electric power plant. It is shown how we can solve such a real world problem using a genetic algorithm (GA) approach together with a fuzzy programming technique. The problem is treated as a two-phase problem. The higher level problem determines the purchase sequence of coals and the lower level problem determines the reception dates of the sequentially coming coal. For the higher level problem, we apply a GA in order to explore the approximately optimal purchase sequence. For the lower level problem, the solution is obtained by applying some simple rules. Compared to a simple random search (RS) and simulated annealing (SA) approaches, the advantages of the proposed GA approach are examined using numerical examples.

2 A Coal Purchase Planning Problem

The electric power plant under consideration is located on the coast of the Sea of Japan. In that plant, the electricity is generated using coal. The coal is imported from several countries, e.g., Australia, America, Canada, China, South Africa and so on. The coal from the docked ship at the pier is directly stored in the 16 silos. No ship can come alongside the pier on the stormy weather days. They have a lot of stormy weather days in winter season. The coal stored for more than 60 days should be moved to another empty silo in order to avoid the spontaneous combustion. Some kinds of coal cannot be used as fuel without being mixed with some others. The fuel inventory control is not an easy task because of these complex settings. In addition to that, the purchase order of coal is based in principle on an annual sales contract, so that a major revision is not accepted. Thus, the annual purchase planning plays an important role for the efficient fuel inventory control.

Under such circumstances, our problem is to find out the optimal annual purchase planning taking into account several constraints as well as a few objectives. The constraints are as follows;

1. There is a seasonally changeable safety stock level for coal: 160,000 t in summer season and 290,000 t in winter season (from November to March).
2. Twenty-eight kinds of coal with different calorific powers are used.
3. Four kinds of coal cannot be used as fuel without being mixed with some others.
4. The coal stored for more than 60 days should be moved to another empty silo in order to avoid the spontaneous combustion.
5. There are 8 possible coal mining countries to import from. The load displacement of a ship and the lead time depend on the country. However, the load displacement is either 30,000 t, 60,000 t or 80,000 t.
6. The annual purchase planning should be suitable for a given annual electricity generation plan.
7. Only 16 silos are available to store the coal, and in each silo, only one kind of coal with the same reception date can be stored. The capacity of a silo is 33,000 t.
8. Only one ship can come alongside the pier in a day.

On the other hand, the objectives are shown as follows;

1. minimize the deviations from a given seasonally changeable target stock level.
2. minimize the deviations from a given target purchase distribution on the coal mining countries.
3. minimize the number of movements of stored coals to another silo.

The annual purchase planning problem can be formulated as an integer or mixed integer programming problem, but the formulated problem becomes a large-scale problem as can be intuitively conjectured. For example, when the daily stock of 16 silos are regarded as a part of the decision variables, the number of decision variables becomes more than 5,000 (taking into consideration only the number of variables of daily stocks, we get $16 \times 365 = 5840$). Thus, it may require a formidable effort as well as a lot of time to solve the problem even if the conventional integer programming techniques are applied.

In this paper, we demonstrate that such a problem can be solved by a two-phase approach using a GA. We decompose the problem into higher and lower level problems. Namely, the higher level problem determines the purchase sequence of coal and the lower level problem determines the reception date of each ordered kind of coal under a given purchase sequence. Note that the order date can be determined by subtracting the lead time from the reception date. The first objective value is constant in the lower level problem. Thus, we can discard the second objective in the lower level problem. Moreover, we assume that the first objective is much more important than the third. To sum up, we consider that the lower level problem is a single objective (the first one) problem with all constraints described above. On the other hand, the higher level problem is just an exploration problem of the purchase sequence without any constraints. Thus, it is easy to apply a GA approach to the higher level problem.

3 Solving the Lower Level Problem

The lower level problem is the determination of the reception date of each ordered kind of coal so as to minimize the deviations from a given target stock. We make it a rule to use coals according to a FIFO (First In First Out) order. The sum of absolute deviations is adopted for the objective function, i.e.,

$$\min_{u(t)} f_1(u) = \sum_{t=1}^{T} |m(t) - q(t|u)|, \tag{1}$$

where t is a discrete time variable, T the end of the duration of concern, i.e., $T = 365$ in our problem, $m(t)$ the given target stock on the t-th day, $u(t)$ the amount of received coal on the t-th day and $q(t|u)$ the final stock on the t-th day under the coal reception scheduling u. $q(t|u)$ can be easily calculated by a recursive equation based on the annual electricity generation plan, the calorific power of the currently used coal and the coal reception scheduling u. Namely, letting $p(t)$ be the amount of consumed coals on the t-th day which can be calculated from the annual electricity generation plan and the calorific power of the currently used coal, we have the following recursive equation;

$$q(t|u) = \begin{cases} q(t-1|u) + u(t) - p(t), & \text{if } t > 0, \\ q_0, & \text{if } t = 0. \end{cases} \tag{2}$$

In this section, we show that an approximately optimal solution to (1) can be obtained through some simple rules. To this end, we assume that changes of the target stock are not greater than the smallest positive amount of received coal, i.e., 30,000 t. To utilize the results under this assumption, we regard a target stock change bigger than 30,000 t, as a sequential change not greater than 30,000 t. Moreover, we assume that the target stock does not change so rapidly. Even if this second assumption is not fulfilled, we may regard the solution obtained from the results under this assumption as an approximately optimal solution.

Now let us discuss the rules which yield an approximately optimal solution to (1). We must consider the constraints described in the preceding section, but, for a while, we discard them for the sake of simplicity. Moreover, for a while, we consider the time variable t continuous so that $f_1(u)$ and $q(t|u)$ are represented as

$$f_1(u) = \int_0^T |m(t) - q(t|u)| dt, \tag{1'}$$

$$\frac{dq(t|u)}{dt} = u(t) - p(t) \tag{2'}$$

The kind and the amount of coal to be received after the coal received at time t^0 have already been determined in the given purchase sequence. Thus, the problem is only to determine the reception date of the forthcoming coal iteratively.

We have three possible cases: (a) the case where the target stock does not change from one reception to the next, (b) the case where the target stock decreases from one reception to the next and (c) the case where the target stock increases from one reception to the next. Let n be the amount of received coal under consideration. In each case, we have the following propositions.

Proposition 1. *Let*

$$u_\tau(t) = \begin{cases} u(t), & \text{if } t \in (0, \tau], \\ 0, & \text{if } t \in (\tau, T]. \end{cases} \tag{3}$$

In case (a), the optimal reception date is t^* which satisfies

$$2(m(t^*) - q(t^*|u_{t^0})) = n. \tag{4}$$

Proof. Let $\hat{t}$ be the optimal reception date. It is obvious that $\hat{t}$ satisfies $t^1 \leq \hat{t} \leq t^2$, where t^1 and t^2 are times such that $q(t^1|u_{t^0}) = m$ and $q(t^2|u_{t^0}) = m - n$, respectively. Let

$$u[\tau](t) = \begin{cases} u(t), & \text{if } t < t^1 \text{ or } t > t^2, \\ n\delta(0), & \text{if } t = \tau, \\ u_{t^0}(t), & \text{otherwise}, \end{cases}$$

where $t^1 \leq \tau \leq t^2$ and δ is the Dirac's distribution. A function $u[\tau]$ is a candidate for the optimal solution to problem (1). $q(t|u[\tau])$ is the stock level when the considered coal is received at time τ. Two extreme functions $q(t|u[t^1])$ and $q(t|u[t^2])$ are illustrated in Fig. 1.

Supposing $t^1 \leq \hat{t} < t^*$, then we have

$$f_1(u[t^*]) - f_1(u[\hat{t}]) = \int_{\hat{t}}^{t^*} |m(t) - q(t|u[t^2])| dt - \int_{\hat{t}}^{t^*} |m(t) - q(t|u[t^1])| dt.$$

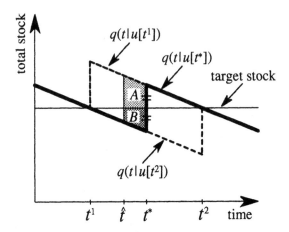

Fig. 1. The constant target stock case

The first term of the right-hand side is depicted as trapezoid B in Fig. 1 while the second term is depicted as trapezoid A in Fig. 1. Since $q(t|u[t^1]) - q(t|u[t^2]) = n$, for all $t \in [t^1, t^2]$, $|m(\hat{t}) - q(\hat{t}|u[t^2])| < |m(\hat{t}) - q(\hat{t}|u[t^1])|$. By the definition of t^*, we obtain $|m(t^*) - q(t^*|u[t^2])| < |m(t^*) - q(t^*|u[t^1])|$. Therefore,

$$\int_{\hat{t}}^{t^*} \big|m(t) - q(t|u[t^2])\big| dt < \int_{\hat{t}}^{t^*} \big|m(t) - q(t|u[t^1])\big| dt,$$

i.e., $f_1(u[t^*]) < f_1(u[\hat{t}])$. This contradicts the optimality of $\hat{t}$. Hence, we have $\hat{t} \geq t^*$.

By the same way, we obtain $\hat{t} \leq t^*$ from the supposition $t^* < \hat{t} \leq t^2$. Hence, $\hat{t} = t^*$. □

Proposition 2. *Let us consider the case where the target stock decreases at the time t^1. If $2(m(t^1) - q(t^1|u_{t^0})) < n$, the optimal reception time is one of the followings:*

1. *time t^2 such that $2(m(t^2) - q(t^2|u_{t^0})) = n$ and $t^2 < t^1$,*
2. *time t^3 such that $2(m(t^3) - q(t^3|u_{t^0})) = n$ and $t^3 > t^1$.*

If $2(m(t^1) - q(t^1|u_{t^0})) \geq n$, the optimal reception time is either t^1 or t^2.

Proof. Let $\hat{t}$ be the optimal time. First we consider the case where $2(m(t^1) - q(t^1|u_{t^0})) < n$. It is obvious that we have $t^2 \leq \hat{t} \leq t^3$. Supposing that $t^2 < \hat{t}$ or $t^2 < \hat{t} < t^1$, we have $f_1(u[t^2]) < f_1(u[\hat{t}])$ by a similar discussion to the proof of Proposition 1 (see Fig. 2). Thus, $\hat{t} = t^2$ or $\hat{t} \geq t^1$. Supposing that $t^1 \leq \hat{t} < t^3$, similarly, we have $f_1(u[t^3]) < f_1(u[\hat{t}])$. Hence, $\hat{t} = t^2$ or $\hat{t} = t^3$.

The proposition in the case where $2(m(t^1) - q(t^1|u_{t^0})) \geq n$ can be proved in the same way. □

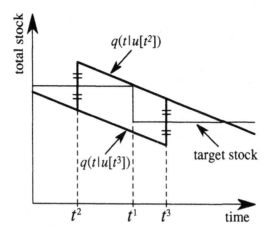

Fig. 2. The decreasing target stock case

Proposition 3. *Let us consider the case where the target stock increases at time* t^1. *If* $2(m(t^1) - q(t^1|u_{t^0})) < n$, *the optimal reception time is time* t^2 *such that* $2(m(t^2) - q(t^2|u_{t^0})) = n$. *Otherwise, the optimal reception time is* t^1.

Proof. It can be proved in the same way as that of Proposition 1. □

Making the time variable discrete again, the following rules are derived from the above propositions.

Rule 1. *When the target stock does not change from one reception to the next, the optimal reception date is* t^* *satisfying*

$$2(m(t^* - 1) - q(t^* - 1|u_{t^0})) < n \leq 2(m(t^*) - q(t^*|u_{t^0})). \qquad (5)$$

Rule 2. *When the target stock decreases at the* t^1-*th day, if* $2(m(t^1) - q(t^1|u_{t^0})) < n$, *the optimal reception date is one of the followings:*

1. *the* t^2-*th day such that* $2(m(t^2 - 1) - q(t^2 - 1|u_{t^0})) < n \leq 2(m(t^2) - q(t^2|u_{t^0}))$ *and* $t^2 < t^1$,
2. *the* t^3-*th day such that* $2(m(t^3 - 1) - q(t^3 - 1|u_{t^0})) < n \leq 2(m(t^3) - q(t^3|u_{t^0}))$ *and* $t^3 > t^1$.

If $2(m(t^1) - q(t^1|u_{t^0})) \geq n$, *the optimal reception date is either* t^1-*th day or* t^2-*th day.*

Rule 3. *When the target stock increases at the* t^1-*th day, if* $2(m(t^1) - q(t^1|u_{t^0})) < n$, *the optimal reception date is the* t^2-*th day such that* $2(m(t^2 - 1) - q(t^2 - 1|u_{t^0})) < n \leq 2(m(t^2) - q(t^2|u_{t^0}))$ *and* $t^2 < t^1$. *Otherwise, the optimal reception date is the* t^1-*th day.*

In order to know which rule is to be used, it is enough to calculate t^* of (5) and the smallest t^{**} such that $n_2 \leq 2(m(t^{**}) - q(t^{**}|u_{t^0}))$ under the assumption that the n tons of coal under consideration is received on the t^*-th day, where n_2 is the amount of the coal to be received next. Namely, if $m(t^*) = m(t^{**})$, Rule 1 is applicable, if $m(t^*) > m(t^{**})$, Rule 2 is applicable, and if $m(t^*) < m(t^{**})$, Rule 3 is applicable.

Rules 1 and 3 give a certain reception date but Rule 2 does not. To know the exact reception date from the result of Rule 2, we calculate the sum of deviations until the t^1- or t^3-th day in both cases and select the date yielding the smallest sum of deviations.

Now let us introduce the constraints to the results of Rules 1, 2 and 3. Let t^* be the optimal reception date obtained from those rules and $s(t)$ be the safety stock on the t-th day. If $q(t^*|u_{t^*}) < s(t^*)$, t^* is changed to $t^{**} = t^* - \delta$, where $\delta > 0$ is the smallest number such that $q(t^* - \delta|u_{t^*}) \geq s(t^* - \delta)$. If there is no empty silo to which the stored coal should move, the last amount of coal scheduled to be received at the time of this conflict is to be postponed until the next day after moving the stored coal. If there are no empty silos large enough to store the coal to be received, the reception date is postponed. The last two modification rules are seldom fired in the real world setting. Thus, by those modifications, we can obtain the approximately optimal reception days.

4 Solving the Higher Level Problem

As shown in the preceding section, we can obtain the approximately optimal reception date of each ordered amount of coal under a given annual purchase sequence. The next problem is to find good annual purchase sequence. However, at this stage, we must consider three objective functions, i.e., the deviations from a given seasonally changeable target stock, the deviations from a given target purchase distribution on the coal mining countries and the number of movements of stored coals from one silo to another. Since the first objective has already been formulated as $f_1(u)$ in (1), let us work on formulating the second one. Let $r = (r_1, r_2, \ldots, r_k)$ be an annual order sequence, where r_j is a code which stands for the kind and the amount of coal. Each kind of coal comes from one mining country. Let $w(r_j, i)$ be a country indicator function such that $w(r_j, i) = 1$ if the code r_j implies importing from the i-th country and $w(r_j, i) = 0$ otherwise. Let $\nu(r_j)$ be the amount of coal corresponding to the code r_j. Moreover, let $d = (d_1, d_2, \ldots, d_v)$ be a distribution on the coal mining countries. Thus, the second objective function can be formulated as

$$f_2(r) = \sqrt{\sum_{i=1}^{v} \left(d_i - \frac{\sum_{j=1}^{k} \nu(r_j)w(r_j,i)}{\sum_{i=1}^{v}\sum_{j=1}^{k} \nu(r_j)w(r_j,i)} \right)^2}. \tag{6}$$

The third objective function value can be calculated while solving the lower level problem. We do not express it in a function form but its value is denoted by

$f_3(r, u)$. Since the first objective function value depends on r as well as u, we prefer to use the notation $f_1(u|r)$ instead of $f_1(u)$.

Since the problem has multiple objective functions (a vector function), it is tractable to scalarize the vector function. To this end, we introduce a fuzzy programming approach [1] [2]. Namely, establishing a decreasing linear membership function μ_{G_i},

$$\mu_{G_i}(v) = \begin{cases} 1, & \text{if } v \leq a_i^1, \\ \dfrac{v - a_i^0}{a_i^1 - a_i^0}, & \text{if } a_i^1 < v \leq a_i^0, \\ 0, & \text{if } v > a_i^1, \end{cases} \tag{7}$$

to each objective function f_i, the three objective functions are aggregated as

$$\max_{r,u} \ \min(\mu_{G_1}(f_1(u|r)), \mu_{G_2}(f_2(r)), \mu_{G_3}(f_3(r, u))). \tag{8}$$

Note that since u is determined in the lower level problem depending on r, (8) can be rewritten as

$$\max_{r} \ \min(\mu_{G_1}(f_1(u^r|r)), \mu_{G_2}(f_2(r)), \mu_{G_3}(f_3(r, u^r))). \tag{9}$$

There are no constraints on the decision variable r, thus, we can easily apply a genetic algorithm to the exploration of the solution of (9). In this paper, a GA [3] [4] is applied and compared to the simple random search and simulated annealing approaches.

In what follows, we describe how a GA is installed.

Coding. A different integer is assigned to each possible pair of kind and amount of the received coal. Thus, a purchase sequence r is represented as a sequence of those integers. Since we do not know the length of the sequence r correctly, we consider a sequence $\hat{r}$ with a sufficiently long length. When the lower level problem is solved, we get a subsequence, r, of $\hat{r}$ whose elements are received by the electric power plant during the duration of concern (one year). Finally, $\hat{r}$ is adopted as an individual representation in the GA.

Crossover. Two individuals are chosen from the population. The chosen individuals are mated with the given crossover probability p_c. Two points crossover is adopted in this paper.

Selection. Introducing the elitist model, the best two individuals survive unconditionally. The other individuals of the next population are chosen based on a ranking selection. We assign the probability mass $p_s(1)$ to the first ranked individual twice as much as the probability mass $p_s(N)$ to the last (N-th) ranked individual, where N is the population size. We produce the arithmetical progression from $p_s(1) = 2p_s(N)$ to $p_s(N)$ so that the j-th ranked individual have the probability mass $p_s(j) = p_s(N)(2N - j - 1)/(N - 1)$. The $p_s(N)$ is determined as $2/3N$ from the equation $\sum_{j=1}^{N} p_s(j) = 1$. Hence, the j-th ranked individual have the selection probability $p_s(j) = 2(2N - j - 1)/3N(N - 1)$.

Table 1. Five kinds of coal possible to purchase

coal no.	load displacement of the ship	calorific power (cal/g)	target purchase rate
0	80,000 t	6950	0.2
1	60,000 t	6800	0.2
2	80,000 t	6800	0.3
3	60,000 t	6520	0.2
4	30,000 t	6700	0.1

Table 2. Target and safety stocks

duration	target stock	safety stock
1st – 19th days	200,000 t	160,000 t
20th – 39th days	230,000 t	160,000 t
40th – 45th days	260,000 t	160,000 t

Mutation. Select an element of the sequence $\hat{r}$. If a uniformly random number is less than the given mutation probability p_m, the element is replaced with a random number obeying the uniform distribution on the coal mining countries.

Initial population. An individual of the initial population is established by repetitively generating random numbers obeying the uniform distribution on the coal mining countries. Repeat this procedure until the initial population is settled.

Fitness function. The objective function (9) is directly adopted for the fitness function fit in GA, i.e.,

$$fit(\boldsymbol{r}) = \min(\mu_{G_1}(f_1(u^r|\boldsymbol{r})), \mu_{G_2}(f_2(\boldsymbol{r})), \mu_{G_3}(f_3(\boldsymbol{r}, u^r))). \qquad (10)$$

5 Numerical Experiments

5.1 Approximation in the Small-sized Problems

For the purpose of checking the proximity of the proposed GA solution to the optimal solution, we consider small-sized problems because the real-world problem is too large to obtain the optimal solution by the complete enumeration. As the small-sized problems, we consider a problem with 45 days duration ($T = 45$), 5 kinds of coal and at most 8 times purchase. Even in this size, we have about 390 thousands (precisely, $5^8 = 390,625$) alternatives including infeasible ones.

Table 3. Initial state of the coal stock in 16 silos in the small-sized problem

silo no.	1	2	3	4	5	6	7	8	9	10	11	12	13	14	15	16
amount (t)	0	0.5	0.5	3.3	3.3	3.3	3.3	3.3	3.3	3.3	0	0	0	0	0	0
kind no.	-	0	0	0	1	1	2	2	3	4	-	-	-	-	-	-
oldness	-	30	30	30	11	12	13	14	15	16	-	-	-	-	-	-

Table 4. The parameters of the membership function for the small-sized problem

	μ_{G_1}	μ_{G_2}	μ_{G_3}
a_i^0	200	200	3
a_i^1	0	0	1

We assume the load displacement of the available ship, the calorific power and the target purchase rate of each coal given in Table 1 and target and safety stocks given in Table 2. All kinds of coal can be used as fuel without being mixed with some others. 100% output is assigned to every day in the generating plan. The initial state of the fuel stock is shown in Table 3. The membership functions μ_{G_i}'s are established by the parameters a_i^0's and a_i^1's shown in Table 4. By the complete enumeration, we obtain 0.702631 as the optimal fitness function value. There are 360,386 feasible solutions. The frequency distribution table is given in Table 5.

We applied the proposed GAs to this problem varying the crossover probability p_c and the mutation probability p_m as in Table 6. We calculated 100 generations with a population of 100 for each simulation. We did 5 simulations

Table 5. Frequency distribution table of the solutions

$fit(r)$	frequency	frequency of GA solutions
0.68–0.71	1116	30
0.65–0.68	20688	0
0.62–0.65	72031	0
0.59–0.62	118763	0
0.56–0.59	101643	0
0.53–0.56	37570	0
less than 0.53	8575	0

Table 6. The crossover and the mutation probabilities

	case 1	case 2	case 3	case 4	case 5	case 6
p_c	0.4	0.5	0.6	0.4	0.5	0.6
p_m	0.05	0.05	0.05	0.1	0.1	0.1

Table 7. The results of simulations in the small-sized problem

	maximum	minimum	average	variance
case 1	0.702631	0.697934	0.700644	0.000004
case 2	0.702631	0.702631	0.702631	0.000000
case 3	0.702631	0.702631	0.702631	0.000000
case 4	0.702631	0.700013	0.702107	0.000001
case 5	0.702631	0.687725	0.699044	0.000041
case 6	0.702631	0.700013	0.701584	0.000002

for each case in Table 6. The maximum, the minimum, the average and the variance of the fitness values are shown in Table 7. The locations of the obtained 30 GA solutions in the frequency distribution table are shown in Table 5. From these, we can recognize that good approximate solutions are obtained with a high probability by the proposed GA approach. Indeed, 21 GA solutions were reached to the optimum. Furthermore, we did similar simulations for different small-sized problems and got similar results.

5.2 Comparison with Other Approaches in the Real-world Problem

We are now in position to apply the proposed method to the coal purchase planning for a real electric power plant. In order to see the usefulness of the genetic operations, we compare the solution obtained by the proposed method with those by a simple random search as well as by simulated annealing. In the simple random search, we generate annual purchase sequences based on the uniform distribution on the coal mining countries, evaluate each annual purchase sequence and take the best one. In the simulated annealing, we generate an annual purchase sequence (solution) r^0 and replace it to a neighboring solution

Table 8. The annual generating plan

month	Jan.	Feb.	Mar.	Apr.	May	Jun.	Jul.	Aug.	Sep.	Oct.	Nov.	Dec.
out put (%)	70	70	70	70	70	70	82	100	100	70	70	70

Table 9. Initial state of the coal stock in 16 silos

silo no.	1	2	3	4	5	6	7	8	9	10	11	12	13	14	15	16
amount (t)	0	0.5	0.5	3.3	3.3	3.3	3.3	3.3	3.3	3.3	3.3	3.3	0	0	0	0
kind no.	-	2	3	4	5	6	7	8	9	10	11	12	-	-	-	-
oldness	-	50	50	60	11	12	13	14	15	16	17	18	-	-	-	-

Table 10. The target annual purchase distribution

class no.	1	2	3	4	5
rate (%)	20	5	50	20	5

r^1 with the transition probability p_t,

$$p_t = \min \left(\exp \left(\frac{fit(r^1) - fit(r^0)}{T} \right), 1 \right),$$ (11)

where T is a temperature parameter decreasing with the iteration number. Every 100 iterations, T is updated by

$$T_{\text{new}} = 0.9 \times T_{\text{old}}.$$ (12)

This updating procedure is repeated a number of times. The initial temperature parameter is set at $T = 10,000$. To obtain a neighboring solution, we generate a uniformly random number R from $\{1, 2, 3\}$ and R randomly selected purchase orders from the given annual purchase sequence are replaced with randomly generated purchase orders.

The circumstances of this problem are as follows. The annual generating plan is given in Table 8. We have 48 pairs of kind and amount of received coal. The calorific power of the coal is from around 6000 to around 7000 (cal/g). Five pairs are not available for the fuel without mixing them with others. Table 3 shows the initial state of the fuel stock. Eight coal mining countries are divided into 5 classes. The target annual purchase distribution on 5 classes are given in Table 10. Table 11 shows the target coal stock.

Table 11. The target coal stock ($\times$ 1,000 t)

duration	Jan.1–Mar.31	Apr.1–Sep.14	Sep.15–30
amount	320	170	200

duration	Oct.1–9	Oct.10–19	Oct.20–31	Nov.1–Dec.31
amount	230	260	290	320

Table 12. The parameters of the membership functions

	μ_{G_1}	μ_{G_2}	μ_{G_3}
a_i^0	1000	510	3
a_i^1	600	10	1

Table 13. The results of simulations

	maximum	minimum	average	variance
GA (case 1)	0.829893	0.717630	0.765263	0.001948
GA (case 2)	0.806410	0.666900	0.732066	0.002246
GA (case 3)	0.902075	0.733172	0.802248	0.004232
GA (case 4)	0.867661	0.696363	0.777420	0.003458
GA (case 5)	0.824722	0.713813	0.778193	0.001496
GA (case 6)	0.789683	0.708789	0.759833	0.000589
RS (20,000)	0.601102	0.505285	0.546607	0.001138
SA (10,000)	0.843110	0.500000	0.661482	0.024794
SA (15,000)	0.867825	0.500000	0.731453	0.019575
SA (20,000)	0.886844	0.500000	0.774443	0.016602

Under this setting, the proposed GA approach, the RS approach and the SA approach are applied to the coal purchase planning problem. The crossover probability p_c and the mutation probability p_m are set as in Table 6 for GA. The membership functions μ_{G_i}'s are established by the parameters a_i^0's and a_i^1's shown in Table 12. We calculate 100 generations with a population of 100; thus, 10,000 individuals are explored. On the other hand, 20,000 individuals are explored in the simple random search and the procedure is independently repeated 10,000, 15,000 and 20,000 times in the simulated annealing. Since those three approaches involve the probabilistic search, we did 10 simulations for each and calculated the maximum, the minimum, the average and the variance of the fitness function value. The results are shown in Table 13. The total stock transitions obtained from the best GA solution and from RS solution are depicted in Fig. 3 and 4, respectively. The obtained purchase rates of the best GA and RS solutions are depicted in Fig. 5.

As shown in Table 13, even the worst solution by GA is better than the best solution by the random search. Comparing the best solution by the GA to that by the RS in three membership values shown in Fig. 3 and 4, we can see a great difference in the first membership values $\mu_{G_1}(f_1(u|r))$. It seems that the membership function of the first fuzzy goal is the tightest among the three. In our current problem setting, GA produces a much better solution than the simple random search.

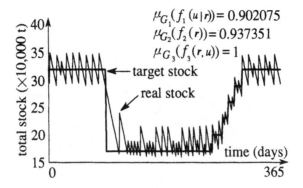

Fig. 3. Total stock transition of the best GA solution

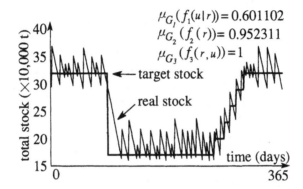

Fig. 4. Total stock transition of the best RS solution

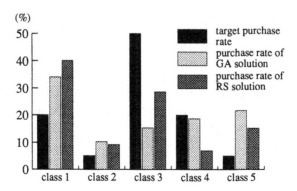

Fig. 5. The obtained purchase rates

Table 14. The results of simulations with a stronger fuzzy goal

	maximum	minimum	average	variance
GA (case 1)	0.828671	0.726253	0.784313	0.001575
GA (case 2)	0.797304	0.689210	0.749093	0.001534
GA (case 3)	0.829948	0.723599	0.778575	0.001413
GA (case 4)	0.803866	0.722711	0.758603	0.000817
GA (case 5)	0.822901	0.732925	0.770503	0.000800
GA (case 6)	0.782130	0.655434	0.719080	0.001673
RS (20,000)	0.563566	0.516009	0.541445	0.000265
SA (10,000)	0.856172	0.500000	0.741324	0.011614
SA (15,000)	0.879509	0.638838	0.784625	0.004611
SA (20,000)	0.836762	0.500000	0.707110	0.024167

Compared to GA solutions, the obtained maximum fitness function values of SA are sometimes better. However, the minimum fitness function values of SA in the cases of 10,000 and 15,000 repetitions are the worst. Moreover, the variance of those cases are larger than the others. Thus, it can be conjectured that the SA solutions significantly depend on the initial solution. The employed SA approach does not have good convergence property to the optimum. On the other hand, the proposed GA approach seems to have a better convergence property than the employed SA approach.

As mentioned before, the first fuzzy goal can be the tightest among the three. In this second set of experiments, we tighten the second fuzzy goal so that a_2^0 is changed from 510 to 110. We ran the same simulations and got the results in Table 14. The total stock transitions obtained from the best GA solution and RS solution are depicted in Fig. 6 and 7, respectively. The obtained purchase rates of the best GA and RS solutions are depicted in Fig. 8.

From Table 14, we can draw a similar discussion to that of Table 13. Note that the minimum fitness function value of SA solutions with 20,000 repetitions is the worst. This shows the weak point of the employed SA approach. Comparing Fig. 6 and 7 with Fig. 3 and 4, we recognize that the total deviations from the target stocks become worse because of the tightened second fuzzy goal. Comparing Fig. 8 with Fig. 5, we recognize that the total deviations from the target purchase distribution are improved. This means that the decision maker's preference can be reflected in the solution through the employed fuzzy programming technique. In consequence, the GA approach stably produces a better solution than the RS and SA approaches in our current problem setting.

6   Concluding Remarks

A two-phase approach using a genetic algorithm and a fuzzy programming technique has been proposed to coal purchase planning in electric power plants. In

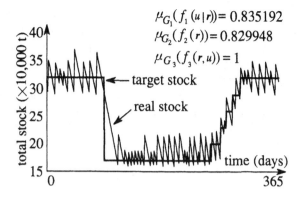

Fig. 6. Total stock transition of the best GA solution for the strongest fuzzy goal

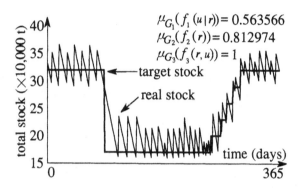

Fig. 7. Total stock transition of the best RS solution for the strongest fuzzy goal

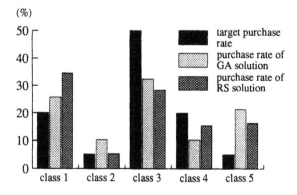

Fig. 8. The obtained purchase rates for the strongest fuzzy goal

this approach, the coal purchase planning problem is treated as a two-level problem. In the higher level problem, by applying a few rules to minimize the total deviations from the target stocks, the purchase sequence of the coal is determined. On the other hand, in the lower level problem, the reception dates of the sequentially coming coal are determined by applying a genetic algorithm. By the numerical simulation of a small-sized problem, we have examined how much the GA solution is close to the optimum and then confirmed the soundness of the GA to this problem. Moreover, the proposed GA approach has been applied to a real-world problem and compared to the simple random search and the simulated annealing approaches. Consequently, it is shown that the GA approach stably produces a better solution than the RS and SA approaches in our current problem setting.

A prototype software based on the proposed approach has been encoded and is being tested by domain experts for a future installation at the plant. On the other hand, similar planning problems are found in [5] [6] but they are in the ironworks field. Our proposed approach would be useful for these problems too.

References

1. Inuiguchi, M., Ichihashi, H., Tanaka, H.: Fuzzy programming: a survey of recent developments, in: Słowinski, R., Teghem, J. (eds.): *Stochastic versus Fuzzy Approaches to Multiobjective Programming under Uncertainty*, Kluwer Academic Publishers, Dordrecht (1990) pp. 45–68
2. Sakawa, M. *Fuzzy Sets and Interactive Multiobjective Optimization*, Plenum Press, New York and London (1993)
3. Goldberg, D. E. *Genetic Algorithms in Search, Optimization & Machine Learning.* Addison-Wesley, Reading, Massachusetts (1983)
4. Michalewicz, Z. *Genetic Algorithms + Data Structure = Evolution Programs, Second extended edition.* Springer-Verlag, Berlin (1994)
5. Hanai, H.: Application of linear programming to material purchase, *Communications of the Operations Research Society of Japan*, **31** (1986) pp. 360–364 (in Japanese)
6. Fukumura, S., Sanou, K., Yamakawa, H.: Expert systems for a shipping plan of steel materials and the branch and bound method, *Communications of the Operations Research Society of Japan*, **33** (1988) pp. 33–39 (in Japanese)

"CAM-Brain"
ATR's Billion Neuron Artificial Brain Project
A Three Year Progress Report

Hugo de Garis

Brain Builder Group, Evolutionary Systems Department,
ATR Human Information Processing Research Laboratories,
2-2 Hikaridai, Seika-cho, Soraku-gun, Kansai Science City, Kyoto, 619-02, Japan.
tel. + 81 7749 5 1079, fax. + 81 7749 5 1008, degaris@hip.atr.co.jp
http://www.hip.atr.co.jp/~degaris

Abstract

This work reports on progress made in the first 3 years of ATR's "CAM-Brain" Project, which aims to use "evolutionary engineering" techniques to build/grow/evolve a RAM-and-cellular-automata based artificial brain consisting of thousands of interconnected neural network modules inside special hardware such as MIT's Cellular Automata Machine "CAM-8", or NTT's Content Addressable Memory System "CAM-System". The states of a billion (later a trillion) 3D cellular automata cells, and millions of cellular automata rules which govern their state changes, can be stored relatively cheaply in giga(tera)bytes of RAM. After 3 years work, the CA rules are almost ready. MIT's "CAM-8" (essentially a serial device) can update 200 million CA cells a second. It is likely that NTT's "CAM-System" (essentially a massively parallel device) will be able to update a *hundred billion* CA cells a second. Hence all the ingredients will soon be ready to create a revolutionary new technology which will allow thousands of evolved neural network modules to be assembled into artificial brains. This in turn will probably create not only a new research field, but hopefully a whole new industry, namely "brain building". Building artificial brains with a billion neurons is the aim of ATR's 8 year "CAM-Brain" research project, ending in 2001.

Keywords

Evolutionary Engineering, Artificial Brains, Neural Networks, Genetic Algorithms, Cellular Automata, Cellular Automata Machines (CAMs), Nano-Electronics, Darwin Machines.

1. Introduction

ATR's CAM-Brain project resulted from the experience of the author's thesis work, in which he evolved neural net modules (using concatenated bit-string weights) to control the behavior of a simulated quadruped called "LIZZY", which could walk straight, turn left, turn right, peck at food and mate [de Garis 1994]. Each of these behaviors was controlled by the time varying outputs of a single evolved neural network module, and applied to the angles of the leg components of LIZZY. (As far as he is aware, the author was the first person to evolve neural net dynamics [de Garis 1991], (in the form of walking stick-legs "Walker")). Switching between behaviors involved taking the outputs from one neural net module and feeding them into the inputs of the next module. The next step was to evolve neural net detectors, e.g. for frequency, signal strength, signal strength difference, etc. Finally, neural net "production rule" modules were evolved which could map conditional inputs from detectors to output behaviors. Thus an "intelligent" artificial creature was built, which could detect prey, mates and predators, and then approach and eat or mate, or turn away and flee.

Virtually every neural net that the author tried to evolve, evolved successfully. *The evolution of these fully connected neural network modules proved to be a very powerful technique.* This success made a deep impression on the author, reinforcing his dream of being able to build much more complex artificial nervous systems, even artificial brains. However, every time the author added a neural net module to the Lizzy simulation, its speed on the screen was slowed (on a Mac II computer). Gradually, the necessity dawned on the author that some kind of evolvable hardware solution [de Garis 1993] would be needed to evolve large numbers of neural net modules and at great speed (i.e. electronic speed) in special machines the author calls "Darwin Machines" [de Garis 1993]. Evolving artificial brains directly in hardware remains the ultimate future goal of the author, but in the meantime (since the field of evolvable hardware (EHW, E-Hard) is today only in its infancy), the author compromises by using cellular automata to grow/evolve neural nets in large numbers in RAM, which is cheap and plentiful. (In a year or so, it will be quite possible to have a gigabyte of RAM in one's work-station). By using cellular automata based neural nets which grow and evolve in gigabytes of RAM, it should be possible to evolve large numbers (thousands) of neural net modules, and then assemble them (or even evolve their interconnections) to build an artificial brain. The bottleneck is the speed of the processor which updates the CA cells. State of the art in such processors is MIT's "CAM-8" machine, which can update 200,000,000 CA cells a second.

Very recently, it has been suggested by the author's ATR colleague Hemmi, that NTT's Content Addressable Memory System (CAM-System) might be able to update CA cells at a rate *thousands* of times faster than the MIT machine, i.e. at a *hundred billion* CA cells per second. NTT's machine is massively parallel. Hemmi and his programmer assistant Yoshikawa are now (December 1995) busily engaged in writing software to convert the author's CA rules (in 2D form) into Boolean expressions suitable for the NTT machine. If they succeed in applying this machine to CAM-Brain, then a new era of brain building can begin, because the ability to evolve thousands of neural net modules would become realistic and very practical (for example, to evolve a neural net module inside a cubic space of a million CA cells, i.e. 100 cells on a side, at a hundred billion cells a second, would take at most about 500 clock cycles, i.e. about five milliseconds. *So the evolution of a population of 100 chromosomes over 100 generations could all be done in about one minute.*) All the essential ingredients for brain building would be available (lots of RAM, the CA rules, and fast CA processors). Even if Hemmi does not succeed, then a new machine can be designed to be thousands of times faster than the CAM-8 machine. The author believes the CAM-Brain breakthrough is either less than a year away, or at most only a few years away (the time necessary to design and build a "Super-CAM" machine, probably with the help of NTT).

The above gives an overview of the CAM-Brain research project. What now follows is a more detailed description of CAM-Brain, showing how one grows and evolves CA based neural net modules in 2D and 3D. We begin with the essential idea. Imagine a 2D CA trail which is 3 cells wide (e.g. Fig. 2). Down the middle of the trail, send growth signals. When a growth signal hits the end of the trail, it makes the trail extend, or turn left, or right, or split etc., depending upon the nature of the signal (e.g. see Figs. 3-6). It was the author who hand coded the CA rules which make these extensions, turns, splits etc. happen. The CA rules themselves are *not* evolved. It is the *sequence* of these signals (fed continuously over time into an initialized short trail) that is evolved. This sequence of growth signals is the "chromosome" of a genetic algorithm, and it is this sequence that maps to a cellular automata network. When trails collide, they can form "synapses" (e.g. see Fig. 7). Once the CA network has been formed in the initial "growth phase", it is later used in a second "neural signaling phase". Neural signals move along CA-based axons and dendrites, and across synapses etc. The CA network is made to behave like a conventional artificial neural network (see Fig. 11). The outputs of some of the neurons of the complex recurrent networks which result can be used to control complex time dependent behaviors whose fitnesses can be measured. These fitness values can be used to drive the evolution. By growing/evolving thousands of neural net modules and their interconnections in an incremental evolutionary way, it will be possible to build artificial brains. According to the CAM developers at MIT, it is likely that the next generation of CAMs will achieve an increase in performance of the order of thousands, within 5 years. However, to be able to evolve a billion neuron artificial brain by 2001 (ATR's goal), it is likely that a "nano-CAM" machine (i.e. one which uses nano-scale electronic speeds and densities) will need to be developed. To this end, we are collaborating with an NTT researcher who has developed a nanoscale electronics device, who wants to combine huge numbers of them to behave as nano-scale cellular automata machines.

In the summer of 1994, a two dimensional CAM-Brain simulation was completed which required 11,000 hand crafted CA state transition rules. It was successfully applied to the evolution of maximizing the number of synapses, outputting an arbitrary constant neural signal value, outputting a sine wave of a desired arbitrary period and amplitude and to the evolution of a simple artificial retina which could output the vector velocity of a "white line" which "moved" across an array of "detector" neurons. Work on the 3D simulation should be completed by early 1996, and is expected to take about 150,000 hand crafted CA rules. The Brain Builder Group of ATR took possession of one of MIT's CAM8 machines in the fall of 1994. At the time of writing (December 1995) the porting of the 2D rules from a Sparc20 workstation to the CAM8 is nearing completion. If the porting of the rules of the 3D simulation to this machine is not possible, then a "SuperCAM" machine will be designed specifically for CAM-Brain, with the collaboration of the Evolutionary Technologies (ET) group of NTT, with whom our Brain Builder group of ATR's Evolutionary Systems (ES) group, collaborates closely. The complexity of CAM-Brain will make it largely undesignable, so a (directed) evolutionary approach called "evolutionary engineering" is being used. Neural networks based on cellular automata [Codd 1968], can be grown and evolved at electronic speeds inside state of the art cellular automata machines, e.g. MIT's "CAM8" machine, which can update 200 million cells per second [Toffoli & Margolus 1990]. Since RAM is cheap, gigabytes of RAM can be used to store the states of the CA cells used to grow the neural networks. CA based neural net modules are evolved in a two phase process. Three cell wide CA trails are grown by sending a sequence of growth signals (extend, turn left, turn right, fork left, fork right, T fork) down the middle of the trail. When an instruction hits the end of the trail it executes its function. This sequence of growth instructions is treated as a chromosome in a Genetic Algorithm [Goldberg 1989] and is evolved. Once gigabytes of RAM and electronic evolutionary speeds can be used, genuine brain building, involving millions and later billions of artificial neurons, becomes realistic, and should become concrete within a year or two. The CAM-Brain Project should revolutionize the fields of neural networks and artificial life, and in time help create a new specialty called "Brain Building", with its own conferences and journals.

This work consists of the following sections. Section 2 describes briefly the idea of "Evolutionary Engineering", of which the CAM-Brain Project is an example. Section 3 describes how neural networks can be based on cellular automata [Codd 1968], and evolved at electronic speeds. Section 4 presents some of the details of CAM-Brain's implementation. Section 5 shows how using cellular automata machines will enable millions of artificial neural circuits to be evolved to form an artificial brain. Section 6 discusses changes needed for the 3D version of CAM-Brain. Section 7 deals with recent work. Section 8 deals with future work and section 9 summarises.

2. Evolutionary Engineering

Evolutionary Engineering is defined to be *"the art of using evolutionary algorithms (such as genetic algorithms [Goldberg 1989]) to build complex systems."* This work reports on the idea of evolving cellular automata based neural networks at electronic speeds inside cellular automata machines. This idea is a clear example of evolutionary engineering. Evolutionary engineering will be increasingly needed in the future as the number of components in systems grows to gargantuan levels. Today's nano-electronics for example, is researching single electron transistors (SETs) and quantum dots. Probably within a decade or so, humanity will have full blown nanotechnology (molecular scale engineering), which will produce systems with a trillion trillion components [Drexler 1992]. The potential complexities of such systems will be so huge, that designing them will become increasingly impossible. However, what is too complex to be humanly designable, might still be buildable, as this work will show. By using evolutionary techniques (i.e. evolutionary engineering), it is often still possible to *build* a complex system, even though one does not understand how it functions. This arises from the notion of the "complexity independence" of evolutionary algorithms, i.e. so long as the (scalar) fitness values which drive the evolution keep increasing, the internal complexity of the evolving system is irrelevant. This means that it is possible to successfully evolve systems which function as desired, but which are too complex to be designable. The author believes that this simple idea (i.e. the complexity independence of evolutionary algorithms) will form the basis of most 21st century technologies (dominated by nanotechnology [Drexler 1992]). Thus, evolutionary engineering can "extend the barrier of the buildable", but may not be good science, because its products tend to be black boxes. However, confronted with the complexity of trillion trillion component systems, evolutionary engineering may be the only viable method to build them.

3.   Cellular Automata Based Neural Networks

Building an artificial brain containing billions of artificial neurons is probably too complex a task to be humanly designable. The author felt that brain building would be a suitable task for the application of evolutionary engineering techniques. The key ideas are the following. Use evolutionary techniques to evolve neural circuits in some electronic medium, so as to take advantage of electronic speeds. The medium chosen by the author was that of cellular automata (CA) [Codd 1968], using special machines, called "Cellular Automata Machines (CAMs)", which can update hundreds of millions of CA cells a second [Toffoli & Margolus 1990].

CAMs can be used to evolve the CA based neural networks at electronic speeds. The states of the cellular automata cells can be stored in RAM, which is cheap, so one can have gigabytes of RAM to store the states of billions of CA cells. This space is large enough to contain an artificial brain. MIT's Information Mechanics Group (Toffoli and Margolus) believe that within a few years it will be technically possible to update a trillion CA cells in about 0.1 nanoseconds [p221, Toffoli & Margolus 1990]. Thus, if CA state transition rules can be found to make CA behave like neural networks, and if such CA based networks prove to be readily evolvable, then a potentially revolutionary new technology becomes possible. The CAM-Brain Project is based on the above ideas and fully intends to build artificial brains before the completion of the project in 2001. The potential is felt to be so great that it is likely that a new specialty will be formed, called "Brain Building".

For the first 18 months of the CAM-Brain Project, the author simulated a two dimensional version of CAM-Brain on a Sparc 10 workstation. This work was completed in the summer of 1994. The 2D version was used briefly (before work on the 3D version was started) to undertake some evolutionary tests, whose results will be presented in the next section. The 2D version served only as a feasibility and educational device. Since trails are obliged to collide in 2D, the 2D version was not taken very seriously. Work was begun rather quickly on the more interesting 3D version almost immediately after the 2D version was ready. Proper evolutionary tests will be undertaken once the 3D version is ready, which should be by early 1996. To begin to understand how cellular automata [Codd 1968] can be used as the basis for the growth and evolution of neural networks, consider Fig. 1 which shows an example of a 2D CA state transition rule, and Fig. 2 which shows a 2D CA trail, 3 cells wide. All cells in a CA system update the state of their cells synchronously. The new state of a given cell depends upon its present state and the states of its nearest neighbors. Down the middle of the 3 cell wide CA trail, move "signal or growth cells" as shown in Fig. 2 As an example of a state transition rule which makes a signal cell move to the right one square, consider the right hand most signal cell in Fig. 2, which has a state of 5. The cell immediately to its right has a state of 1, which we want to become a 5. Therefore the 2D state transition rule to turn the 1 into a 5 is 1.2.2.2.5-->5. These signal or growth cells are used to generate the CA trails, by causing them to extend, turn left or right, split left or right, and Tsplit. When trails collide, they can form synapses. It is the sequence of these signal cells which determines the configuration of the CA trails, thus forming a CA network. It is these CA trails which later are used as neural network trails of axons and dendrites. Neural signals are sent down the middle of these CA trails. Thus there are two major phases in this process. Firstly, the CA trails are grown, using the sequence of signal cells. Secondly, the resulting CA trail network is used as a neural network, whose fitness at controlling some system can be measured and used to evolve the original growth sequence. To make this more explicit, it is the sequence of growth cells which is evolved. By modifying the sequence, one alters the CA network configuration, and hence the fitness of the configuration when it functions as a neural net in the second phase. From a genetic algorithm (GA) point of view, the format of the GA "chromosome" is the sequence of integers which code for the signaling or growth instructions. By mutating and crossing over these integers, one obtains new CA networks, and hence new neural networks. By performing this growth at electronic speeds in CAMs, and in parallel, with one CAM per GA chromosome, and attaching a conventional programmable microprocessor to each CAM to measure the user defined fitness of the CA based neural circuit, one has a means to evolve large numbers of neural modules very quickly. Using CAMs to evolve neural circuits, is an example of a type of machine that the author labels a "Darwin Machine", i.e. one which evolves its own structure or architecture. A related idea of the author concerns the concept of "Evolvable Hardware (EHW)" [de Garis 1993] where the software instructions used to configure programmable logic devices (PLDs) are treated as chromosomes in a Genetic Algorithm [Goldberg 1989]. One then rewrites the circuit for each chromosome.

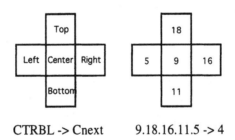

CTRBL -> Cnext   9.18.16.11.5 -> 4

Fig. 1 A 2D CA State Transition Rule

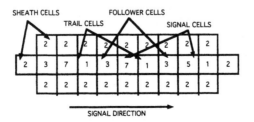

Fig. 2 Signal Cells Move Along a Cellular Automata Trail

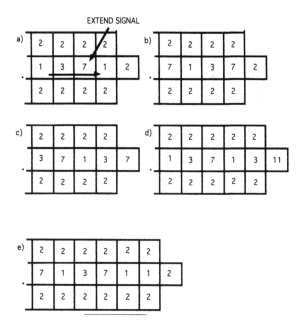

Fig. 3   Extend the Trail

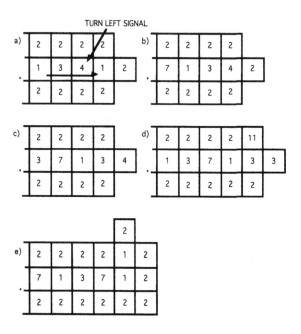

Fig. 4   Turn Trail Left

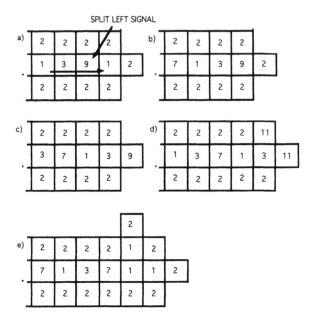

Fig. 5 Split Trail Left

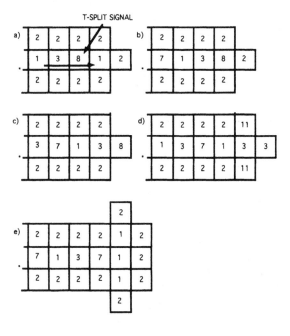

Fig. 6 T-Split Trail

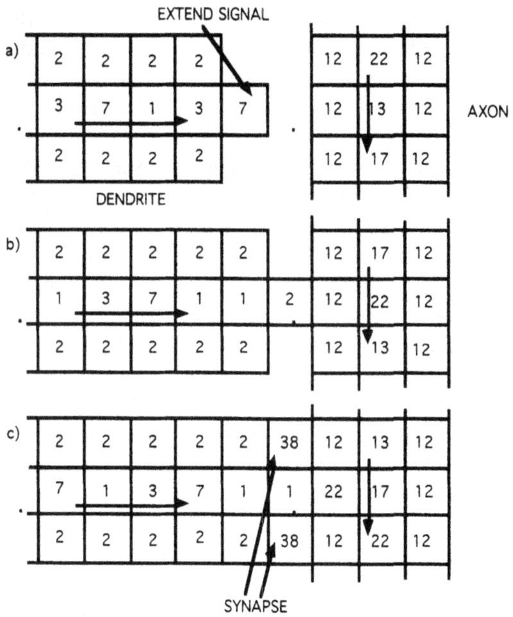

Fig. 7   Dendrite to Axon Synapsing

4.   Further Details

This section provides further details on the implementation of the CA based neural networks. There are three kinds of CA trails in CAM-Brain, labeled dendrites, excitatory axons and inhibitory axons, each with their own states. Whenever an axon collides with a dendrite or vice versa, a "synapse" is formed. When a dendrite hits an excitatory/inhibitory axon or vice versa, an excitatory/inhibitory synapse is formed. An inhibitory synapse reverses the sign of the neural signal value passing through it. An excitatory synapse leaves the sign unchanged. Neural signal values range between -240 and +240 (or their equivalent CA states, ranging from 100 to 580). The value of a neural signal remains unchanged as it moves along an axon, but as soon as it crosses a synapse into a dendrite, the signal value (i.e. signal strength) begins to drop off linearly with the distance it has to travel to its receiving neuron. Hence the signal strength is proportional to the distance between the synapse and the receiving neuron. Thus the reduction in signal strength acts like a weighting of the signal by the time it reaches the neuron. But, this distance is evolvable, hence indirectly, the weighting is evolvable. CAM-Brain is therefore equivalent to a conventional artificial neural network, with its weighted sums of neural signal strengths. However, in CAM-Brain there are time delays, as signals flow through the network. When two or three dendrite signals collide, they sum their signal strengths (within saturated upper or lower bounds).

When implementing the 2D version of CAM-Brain, it soon became noticeable that there were many many ways in which collisions between CA trails could occur. So many, that the author became increasingly discouraged. It looked as though it would take years of handcoding the CA state transition rules to get CAM-Brain to work. The intention was to have rules which would cover every collision possibility. Eventually a decision was made to impose constraints on the ways in which CA trails could grow. The first such constraint was to make the trails grow on a grid 6 cells or squares (cubes) on a side. This process (called "gridding") sharply reduces the number of collision types. It also has a number of positive side effects. One is that in the neural signaling phase, neural signals arrive synchronously at junction points. One no longer needs to have to handcode rules for phase delays in neural signaling summation. By further imposing that different growth cells advance the length of the trails by the same number of squares, one can further reduce the number of collision types. With synchrony of growth and synchrony of signaling and gridding, it is possible to cover all possible types of collisions. Nevertheless, it still took over 11000 rules to achieve this goal, and this was only for the 2D version. The 3D version is expected to take about 150,000 rules, but due to the experience gained in working on the 2D version, and to the creation of certain software productivity tools, the 3D version should be completed by early 1996

Considering the fact that the 2D version takes 11,000 rules, it is impossible in this short work to discuss all the many tricks and strategies that are used to get CAM-Brain to work. That would require a book (something the author is thinking seriously about writing, if he ever makes time to do it). However, some of the tricks will be mentioned here. One is the frequent use of "gating cells", i.e. cells which indicate the direction that dendrite signals should turn at junctions to head towards the receiving neuron. To give these gating cells a directionality, e.g. a "leftness" or a "rightness", special marker cells are circulated at the last minute, after the circuit growth is stabilized. Since some trails are longer than others, a sequence of delay cells are sent through the network after the growth cells and before the marker cells. Without the delay cells, it is possible that the marker cells pass before synapses are formed.

Once the 2D simulation was completed (before the CAM8 was delivered) several brief evolutionary experiments using the 2D version were undertaken. The first, was to see if it would be possible to evolve the number of synapses. Figs. 9, 10, 11 show the results of an elite chromosome evolved to give a large number of synapses. Fig. 9 shows early growth. Fig. 10 shows completed growth, and Fig. 11 shows the neural signaling phase. In this experiment, the number of synapses increased steadily. It evolved successfully. The next experiment was to use the neural signaling to see if an output signal (tapped from the output of one of the neurons) could evolve to give a desired constant value. This evolved perfectly. Next, was to evolve an oscillator of a given arbitrary frequency and amplitude, which did evolve, but slowly (it took a full day on a Sparc10 workstation). Finally, a simple retina was evolved which output the two component directional velocity of a moving "line" which passed (in various directions) over a grid of 16 "retinal neurons". This also evolved but even more slowly. The need for greater speed is obvious.

The above experiments are only the beginning. The author has already evolved (not using CAs) the weights of recurrent neural networks as controllers of an artificial nervous system for a simulated quadrupedal artificial creature. Neural modules called "GenNets" [de Garis 1990, 1991] were evolved to make the creature walk straight, turn left or right, peck at food, and mate. GenNets were also evolved to detect signal frequencies, to generate signal frequencies, to detect signal strengths, and signal strength differences. By using the output of the detector GenNets, it was possible to switch motion behaviors. Each behavior had its own separately evolved GenNet. By switching between a library of GenNets (i.e. their corresponding evolved weights) it was possible to get the artificial creature to behave in interesting ways. It could detect the presence and location of prey, predators and mates and take appropriate action, e.g. orientate, approach, and eat or mate, or turn away and flee. However, every time the author added another GenNet, the motion of the simulated creature slowed on the screen. The author's dream of being able to give a robot kitten some thousand different behaviors using GenNets, could not be realized on a standard monoprocessor workstation. Something more radical would be needed. Hence the motivation behind the CAM-Brain Project.

5. A Billion Neurons in a Trillion Cell CAM by 2001

Fig. 8a shows some estimated evolution times for 10 chromosomes over 100 generations for a Sparc 10 workstation, a CAM8, and a CAM2001 (i.e. a CAM using the anticipated electronics of the year 2001) for a given application. In the 2D version of CAM-Brain, implemented on a Sun Sparc 10 workstation, it takes approximately 3.4 minutes to grow a stable cellular automata network consisting of only four neurons. It takes an additional 3.2 minutes to perform the signaling on the grown network, i.e. a total growth-signaling time to measure the fitness of a chromosome of 6.6 minutes. This time scales linearly with the number of artificial neurons in the network. If one uses a population of 10 chromosomes, for 100 generations, the total evolution time (on a Sparc 10) is 100*10*6.6 minutes, i.e., 110 hours, or 4.6 days. This is obviously tediously slow, hence the need to use a CAM. MIT's CAM8 [Toffoli & Margolus 1990] can update 25 million cellular automata cells per second, per hardware module. A CAM8 "box" (of personal computer size) contains eight such modules, and costs about $40,000. Such boxes can be connected blockwise indefinitely, with a linear increase in processing capacity. Assuming an eight module box, how quickly can the above evolution (i.e. 100 generations, with a population size of 10) be performed? With eight modules, 200 million cell updates per second is possible. If one assumes that the 2D CA space in which the evolution takes place is a square of 100 cells on a side, i.e., 10,000 cells, then all of these cells can be (sequentially) updated by the CAM8 box in 50 microseconds. Assuming 1000 CA clock cycles for the growth and signaling, it will take 50 milliseconds to grow and measure the fitness of one chromosome. With a population of 10, and 100 generations, total CAM8 evolution time for a four neuron network will be 50 seconds, i.e. about one minute, which is roughly 8000 times faster.

Using the same CAM8 box, and a 3D space of a million cells, i.e. a cube of 100 cells on a side, one could place roughly 40 neurons. The evolution time will be 100 times as long with a single CAM8 box. With 10 boxes, each with a separate microprocessor attached, to measure the fitness of the evolved network, the evolution time would be about eight minutes. Thus for 1000 neurons, the evolution would take about 3.5 hours, quite an acceptable figure. For a million neurons, the evolution time would be nearly five months. This is still a workable figure. Note, of course, that these estimates are lower bounds. They do not include the necessary human thinking time, and the time needed for sequential, incremental evolution, etc. However, since the CAM-Brain research project will continue until the year 2001, we can anticipate an improvement in the speed and density of electronics over that period. Assuming a continuation of the historical doubling of electronic component density and speed every two years, then over the next eight years, there will be a 16-fold increase in speed and density. Thus the "CAM-2001" box will be able to update at a rate of 200*16*16 million cells per second. To evolve the million neurons above will take roughly 13.6 hours. Thus to evolve a billion neurons, will take about 19 months, again a workable figure. But, if a million neurons can be successfully evolved, it is likely that considerable interest will be focused upon the CAM-Brain approach, so that more and better machines will be devoted to the task, thus reducing the above 19-month figure. For example, with 100 machines, the figure would be about two months. The above estimates are summarized in Figure 8a. These estimates raise some tantalizing questions. For example, if it is possible to evolve the connections between a billion artificial neurons in a CAM2001, then what would one want to do with such an artificial nervous system (or artificial brain)? Even evolving a thousand neurons raises the same question.

Sparc10	CAM8	CAM8	CAM8	CAM8	CAM2001	CAM2001
10000 CA cells	10000 CA cells	1 million CA cells	25 million CA cells	25 billion CA cells	25 billion CA cells	25 trillion CA cells
4 neurons	4 neurons	40 neurons	1000 neurons	1 million neurons	1 million neurons	1 billion neurons
1 Sparc10	1 CAM8	10 CAM8s	10 CAM8s	10 CAM8s	10 CAM2001s	100 CAM2001s
4.6 days	50 seconds	8 minutes	3.5 hours	5 months	13.6 hours	2 months

Fig. 8a   Evolution Times for Different Machines & CA Cell, Neuron & Machine Numbers

48*48*24	10gens 51	40gens 63	60gens 71	100gens 93	
96*48*24	10gens 81	20gens 89	45gens 122		
96*96*24	5gens 116	10gens 116	40gens 205	45gens 205	70gens 234
96*96*48	5gens 235	10gens 235			

Fig. 8b   Synapses per Neuron Doubles as 3D Space Doubles

One of the aims of the CAM-Brain research project is to build an artificial brain which can control 1000 behaviors of a "robot kitten" (i.e. a robot of size and capacities comparable to a kitten) or to control a household "cleaner robot". Presumably it will not be practical to evolve all these behaviors at once. Most likely they will have to be evolved incrementally, i.e., starting off with a very basic behavioral repertoire and then adding (stepwise) new behaviors. In brain circuitry terms, this means that the new neural modules will have to connect up to already established neural circuits. In practice, one can imagine placing neural bodies (somas) external to the established nervous system and then evolving new axonal and dendral connections to it.

The CAM-Brain Project hopes to create a new tool to enable serious investigation of the new field of "incremental evolution." This field is still rather virgin territory at the time of writing. This incremental evolution could benefit from using embryological ideas. For example, single seeder cells can be positioned in the 3D CA space under evolutionary control. Using handcrafted CA "developmental or embryological" rules, these seeder cells can grow into neurons ready to emit dendrites and axons [de Garis 1992]. The CAM-Brain Project, if successful, should also have a major impact on both the field of neural networks and the electronics industry. The traditional preoccupation of most research papers on neural networks is on analysis, but the complexities of CAM-Brain neural circuits, will make such analysis impractical. However, using Evolutionary Engineering, one can at least build/evolve functional systems. The electronics industry will be given a new paradigm, i.e. evolving/growing circuits, rather than designing them. The long term impact of this idea should be significant, both conceptually and financially.

6.   The 3D Version

The 3D version is a conceptually (but not practically) simple extension of the 2D version. Instead of 4 neighbors, there are 6 (i.e. North, East, West, South, Top, Bottom). Instead of 6 growth instructions as in the 2D version (i.e. extend, turn left, turn right, split extend left, split extend right, split left right), there are 15 in the 3D version. A 3D CA trail cross section consists of a center cell and 4 neighbor cells, each of different state or color (e.g. red, green, blue, brown). Instead of a turn left instruction being used as in the 2D case, a "turn green" instruction is used in the 3D case. The 15 3D growth instructions are (extend, turn red, turn green, turn blue, turn brown, split extend red, split extend green, split extend blue, split extend brown, split red brown, split red blue, split red green, split brown blue, split brown green, split blue green). A 3D CA rule thus consists of 8 integers of the form CTSENWB-->Cnew. The 3D version enables dendrites and axons to grow past each other, and hence reach greater distances. The weakness with the 2D version is that collisions in a plane are inevitable, which causes a crowding effect, whereby an axon or dendrite cannot escape from its local environment. This is not the case with the 3D version, which is topologically quite different. A 3D version is essential if one wants to build artificial brains with many interconnected neural modules. The interconnectivity requires long axons/dendrites. Fig. 12 shows an early result in 3D simulation. A space of 3D CA cells (48*48*48 cubes) was used. A single short 3D CA trail was allowed to grow to saturate the space. One can already sense the potential complexity of the neural circuits that CAM-Brain will be able to build. In 3D, it is likely that each neuron will have hundreds, maybe thousands of synapses, thus making the circuits highly evolvable due to their smooth fitness landscapes (i.e. if you cut one synapse, the effect is minimal when there are hundreds of them per neuron).

7.   Recent Work

Just prior to writing this work, the author was able to test the idea that in 3D a single neuron could have an arbitrarily large number of synapses, provided that there is enough space for them to grow in. This was a crucial test, whose results are shown in Fig. 8b. Fitness was defined as the number of synapses formed for two neurons in CA spaces of 48*48*24, 96*48*24, 96*96*24, and 96*96*48 cells respectively. One can see that by doubling the space, one doubles (roughly) the number of synapses (for the elite chromosome). If this had not been the case, for example, if some kind of fractal effect had caused a crowding of the 3D circuits (similar to the crowding effect in 2D), then the whole CAM-Brain project would have been made doubtful. However, with this result, it looks as though evolvability in the 3D signaling phase will be excellent, although the author needs several months more work before completing the 3D signaling phase to confirm his confidence.

At the time of writing (December 1995), the author is completing the simulation of the 3D version, working on the many thousands of rules necessary to specify the creation of synapses. So far, more than 140,000 3D rules have been implemented, and it is quite probable that the figure may go higher than 150,000. Since each rule is rotated 24 ways (6 ways to place a cube on a surface, then 4 ways to rotate that cube) to cater to all possible orientations of a 3D trail, the actual number of rules placed in the (hashed) rule base will be over 3 million. Specifying these rules takes time, and constitutes so far, the bulk of the effort spent building the CAM-Brain system. Software has been written to help automate this rule generation process, but it remains a very time consuming business. Hence the immediate future work will be to complete the simulation of the 3D version. Probably, this will be done by early 1996.

Early in 1995, the author put his first application on the CAM8 machine (which rests on his desk). MIT's CAM8 is basically a hardware version of a look up table, where the output is a 16 bit word which becomes an address in the look up table at the next clock cycle. This one clock cycle lookup loop is the reason for CAM8's speed. It is possible to give each CA cell in the CAM8 more than 16 bits, but tricks are necessary. The first CAM8 experiment the author undertook involved only 16 bits per CA cell. This work is too short to go into details as to how the CAM8 functions, so only a broad overview will be given here. The 16 bits can be divided into slices, one slice per neighbor cell. These slices can then be "shifted" (by adding a displacement pointer) by arbitrarily large amounts (thus CAM8 CA cells are not restricted to having local neighbors). With only 16 bits, and 4 neighbors in the 2D case (Top, Right, Bottom, Left) and the Center cell, that's only 3 bits per cell (i.e. 8 states, i.e. 8 colors on the display screen). It is not possible to implement CAM-Brain with only 3 bits per CA cell. It was the intention of the author to use the CAM8 to show its potential to evolve neural circuits with a huge number of artificial neurons. The author chose an initial state in the form of a square CA trail with 4 extended edges. As the signals loop around the square, they duplicate at the corners. Thus the infinite looping of 3 kinds of growth signals supply an infinite number of growth signals to a growing CA network. There are 3 growth signals (extend, extend and split left, extend and split right). The structure needs exactly 8 states. The 8 state network grows into the 32 megacells of 16 bits each, which are available in the CAM8. At one pixel per cell, this 2D space takes over 4 square meters of paper poster (hanging on the author's wall). A single artificial neuron can be put into the space of one's little finger nail, thus allowing 25,000 neurons to fit into the space. If 16 Mbit memory chips are used instead of 4 Mbit chips, then the area and the number of neurons quadruples to 100,000.

Placing the poster on the author's wall suddenly gave visitors a sense of what is to come. They could see that soon a methodology will be ready which will allow the growth and evolution of artificial brains, because soon it will be possible to evolve many thousands of neural modules and their inter-connections. The visitors sense the excitement of CAM-Brain's potential.

Filling a space of 32 Mcells, with artificial neurons can be undertaken in at least two ways. One is to use a very large initialization table with position vectors and states. Another, is to allow the neurons to "grow" within the space. The author chose to use this "neuro-embryonic" approach. A single "seeder" CA cell is placed in the space. This seeder cell launches a cell to its right and beneath it. These two launched cells then move in their respective directions, cycling through a few dozen states When the cycle is complete, they deposit a cell which grows into the original artificial neuron shape that the author uses in the 2D version of CAM-Brain. Meanwhile other cells are launched to continue the growth. Thus the 32Mcell space can be filled with artificial neurons ready to accept growth cell "chromosomes" to grow the neural circuitry. This neuro-embryogenetic program (called "CAM-Bryo") was implemented on a workstation by the author, and ported to the CAM8 by his research colleague Felix Gers. In order to achieve the porting, use was made of "subcells" in the CAM8, a trick which allows more than 16 bits per CA cell, but for N subcells of 16 bits, the total CAM8 memory space available for CA states is reduced by a factor of N. Gers used two subcells for CAM-Bryo, hence 16M cells of 32 bits each. A second poster of roughly two square meters was made, which contained about 25,000 artificial neurons (see Fig. 13). Again, with 16Mbit memory chips, this figure would be 100,000. Gers expects to be able to port the 2D version of CAM-Brain to the CAM8 with a few weeks work, in which case, a third poster will be made which will depict about 15,000 neurons (with a lower density, to provide enough space for the neural circuitry to grow) and a mass of complex neural circuits. Once this is accomplished, we expect that the world will sit up and take notice - more on this in the next section.

The author's boss at ATR's Evolutionary Systems department, has recently set up a similar group at his company NTT, called Evolutionary Technologies (ET) department. The idea is that once the ATR Brain Builder group's research principles are fairly solid, the author and the author's boss (whose careers are now closely linked) will be able to tap into the great research and development resources of one of the world's biggest companies, when the time comes to build large scale artificial brains. NTT has literally thousands of researchers.

The author would like to see Japan invest in a major national research project within the next 10 years to build "Japan's Artificial Brain", the so-called "J-Brain Project". This is the goal of the author, and then to see such a project develop into a major industry within 20 years. Every household would like to have a cleaner robot controlled by an artificial brain. The potential market is huge.

8. Future Work

A lot of work remains to be done. The author has a list of "to dos" attached to his computer screen. The first item on the list is of course, to finish the rules for the 3D version of CAM-Brain. This should be done by early 1996, and will probably need over 150,000 CA rules. Second, the experience gained in porting the 700 rules for "CAM-Bryo" from a workstation to the CAM8 will shortly enable Gers to complete the much tougher task of porting the 2D version of CAM-Brain to the CAM8. In theory, since there are 11,000 CA rules for the 2D version, and that each rule has 4 symmetry rotations, that makes about 45,000 rules in total to be ported. This fits into the 64K words addressable by 16 bits. The 3D version however, with its (estimated) 150,000 rules, and its 24 symmetry rotations, will require over 3 million rules in total. The 3D version may require a "Super CAM" to be designed and built (by NTT's "Evolutionary Technologies" Dept., with whom the author collaborates closely), which can handle a much larger number of bits than 16. The group at MIT who built CAM8 is thinking of building a CAM9 with 32 bits. This would be very interesting to the author. Whether NTT or MIT get there first, such a machine may be needed to put the 3D version into a CAM. However, with a state-of-the-art workstation (e.g. a DEC Alpha, which the user has on his desk) and a lot of memory (e.g. 256 Mbyte RAM), it will still be possible to perform some interesting evolutionary experiments in 3D CAM-Brain, but not with the speed of a CAM.

Another possibility for porting the 3D version to the CAM8, is to re implement it using CA rules which are more similar to those used in von Neumann's universal constructor/calculator, rather than Codd's. Von Neumann's 2D trails are only 1 cell wide, whereas Codd's 2D version are 3 cells wide, with the central message trail being surrounded by two sheath cells. The trick to using von Neumann's approach is incorporating the direction of motion of the cell as part of the state. The author's colleague Jacqueline Signorinni advises that CAM-Brain could be implemented at a higher density (i.e. more filled CA cells in the CA space) and without the use of a lookup table. The control of the new states would be implemented far more simply she feels, by simple IF-THEN-ELSE type programming. "von Neumann-izing" the 3D version of CAM-Brain might be a good task for the author's next grad student.

With the benefit of hindsight, if the 3D version is reimplemeted (and it is quite likely that my boss will have other members of our group do just that), then the author would advise the following. If possible (if you are implementing a Codd version) give the four sheath cells in a 3D CA trail cross section the same state. This would obviously simplify the combinatorial explosion of the number of collision cases during synapse formation. But, how then would the 3D growth instructions be interpreted when they hit the end of a trail, and how would you define the symmetry rotations? If possible, it would also be advisable to use the minimum number of gating cell states at growth junctions for all growth instructions. Whether this is possible or not, remains to be seen. However, if these simplifications can be implemented (and of course the author thought of them originally, but was unable to find solutions easily), then it is possible that the number of 3D CA rules might be small enough to be portable to the CAM8, which would allow 3D neural circuits to be evolved at 200 million CA cells per second (actually less because of the subcell phenomenon).

Once the 3D rules are ready, two immediate things need to be done. One is to ask ATR's graphics people to display these 3D neural circuits in an interesting, colorful way, perhaps with VR (virtual reality) 3D goggles with interactivity and zoom, so that viewers can explore regions of the dynamic circuits in all their 500 colors (states). This could be both fun and impressive. The second thing is of course to perform some experiments on the 3D version. As mentioned earlier, this will have to be done on a workstation, until a SuperCAM is built. Another possibility, as mentioned earlier is to redesign the 3D CA rules, to simplify them and reduce their number so that they can fit within the 64K 16 bit confines of the CAM8 machine.

As soon as the 2D rules have been fully ported to the CAM8, experiments can begin at speed. Admittedly the 2D version is topologically different from the 3D version (in the sense that collisions in 2D are easier than in 3D), it will be interesting to try to build up a rather large neural system with a large number of evolved modules (e.g. of the order of a hundred, to start with). At this stage, a host of new questions arise. Look at Fig. 14, which is van Essen's famous diagram of the modular architecture of the monkey's visual and motor cortex, showing how the various geographical regions of the brain (which correspond to the rectangles in the figure, and to distinct signal processing functions) connect with each other. Physiological techniques now exist which enable neuro-anatomists to know which distinct cortical regions connect to others. Thus the geography (or statics) of the biological brain is increasingly known. What remains mysterious of course, is the dynamics. How does the brain function.

Van Essen's diagram is inspirational to the author. The author would like to produce something similar with CAM-Brain, i.e. by evolving neural modules (corresponding to the rectangles, or parts of the rectangles) and their interconnections. This raises other questions about sequencing and control. For example, does one evolve one module and freeze its circuits and then evolve another module, freeze its circuits and finally evolve the connections between them, or does one evolve the two modules together, or what? Will it be necessary to place walls around each module, except for hand crafted I/O trails? The author has no clear answers or experience yet in these matters. The author's philosophy is "first build the tool, and then play with it. The answers will come with using the tool".

Another possibility for future work is to try to simplify the whole process of rule making. Perhaps higher level rules can be made which are far fewer in number and allow the author's low level rules to be generated from them. If such a thing can be done, it would be nice, but the author believes there are still so many special cases in the specification of 3D CAM-Brain, that the number of high level rules may still be substantial. If these high level rules can be found, it might be possible to use them and put them on the CAM8, so that 3D evolutionary experiments can be undertaken at CAM8 speeds. Another idea is to use FPGAs (field programmable gate arrays) which code these high level rules and then to use them to grow 3D neural circuits. Each 3D CA cell could contain pointers to its 3D neighbors. In this way, it would be possible to map 3D neural circuits onto 2D FPGAs. This is longer term work. FPGAs are not cheap if many are needed. The author's RAM based solution has the advantage of being cheap, allowing a billion (one byte) CA cell states to be stored reasonably cheaply.

A recent suggestion coming from NTT concerns the use of an existing "content addressable memory" machine, which may be able to update CA cells effectively. There is a "CAMemory" research group at NTT that ATR is now collaborating with. If a small enough number of CAMemory Boolean function rules corresponding to CAM-Brain can be found (a big if), it is possible that a NTT's CAMemory could be thousands of times faster than the CAM8. Obviously, such a possibility is worth investigating, and if successful, could be extremely exciting, since it would mean *hundreds of billions* of CA cell updates a second.

The author feels that the nature of his research in 1996 will change from one of doing mostly software simulation (i.e. generating masses of CA rules), to learning about the biological brain (i.e. reading about brain science to get ideas to put into CAM-Brain), hardware design, and evolvable hardware. These activities will proceed in parallel. Of course, evolutionary experiments, on CAM8 for the 2D version of CAM-Brain, and on a 256 Mbyte RAM (DEC Alpha) workstation for the 3D version, will also be undertaken in parallel.

Further down the road, will be the attempt to design a "nanoCAM" or "CAM2001" based on nanoelectronics. The Brain Builder Group at ATR is collaborating with an NTT researcher who wants to build nano-scale cellular automata machines. With the experience of designing and building a "SuperCAM", a nanoscale CAM should be buildable with several orders of magnitude greater performance. Further research aims are to use CAs to make Hebbian synapses capable of learning. One can also imagine the generation of artificial "embryos" inside a CA machine, by having CA rules which allow an embryological "unfolding" of cell groups, with differentiation, transportation, death, etc. resulting in a form of neuro-morphogenesis similar to the way in which biological brains are built. The author's "CAM-Bryo" program is an early example of this kind of neuro-morphogenetic research.

9. Summary

The CAM-Brain Project at ATR, Kyoto, Japan, intends to build/grow/evolve a cellular automata based artificial brain of a billion artificial neurons at (nano-)electronic speeds inside Cellular Automata Machines (CAMs) by the year 2001. Quoting from a paper by Margolus and Toffoli of MIT's Information Mechanics group, "We estimate that, with integrated circuit technology, a machine consisting of a trillion cells and having an update cycle of 100 pico-second for the entire space will be technologically feasible within 10 years" (i.e. by 2000) [Margolus and Toffoli 1990]. In a trillion 3D CA cells (cubes), one can place billions of artificial neurons. Such an artificial nervous system will be too complex to be humanly designable, but it may be possible to evolve it, and incrementally, by adding neural modules to an already functional artificial nervous system. In the summer of 1994, a 2D simulation of CAM-Brain using over 11000 hand crafted CA state transition rules was completed, and initial tests showed the new system to be evolvable. By early 1996, a 3D simulation will be completed.

If the CAM-Brain Project is successful, it will revolutionize the field of neural networks and artificial life, because it will provide a powerful new tool to evolve artificial brains with billions of neurons, and at electronic speeds. The CAM-Brain Project will thus produce the first Darwin Machine, i.e. a machine which evolves its own architecture. The author is confident that in time a new specialty will be established, based partly on the ideas behind CAM-Brain. This specialty is called simply "Brain Building".

The author and his colleague Felix Gers are about to port the 2D version of CAM-Brain to the CAM8. Hence in early 1996, it will be possible to evolve neural circuits with 25,000 neurons (or 100,000 neurons, with 16 Mbit memory chips) at 200 million CA cell updates a second. As mentioned earlier, the author expects that when this happens, the world will sit up and take notice. Twenty years from now, the author envisages the brain builder industry (i.e. intelligent robots etc.) as being one of the world's top industries, comparable with oil, automobile, and construction. He sees an analogy between the efforts of the very early rocket pioneers (e.g. the American Goddard, and the German (V2) von Braun) and the US NASA mission to the moon which followed. Today's 100,000-neuron artificial brain is just the beginning of what is to come. With adiabatic (heat generationless) reversible quantum computation, it will be possible to build 3D hardware circuits that do not melt. Hence size becomes no obstacle, which means that one could use planetoid size asteroids to build huge 3D brain like computers containing ten to power 40 components with one bit per atom. Hence late into the 21st century, the author predicts that human beings will be confronted with the "artilect" (artificial intellect) with a brain vastly superior to the human brain with its pitiful trillion neurons. The issue of "species dominance" will dominate global politics late next century. The middle term prospects of brain building are exciting, but long term they are terrifying. The author has written an essay on this question [de Garis 1995]. If you would like to be sent a copy, just email him at degaris@hip.atr.co.jp (The author will set up his home page on the web in 1996, after making the effort to learn html).

Finally, by way of a postscript - as the author was preparing the final draft, there were 6 people at ATR working on CAM-Brain (the author (3D CA rules), and his colleague Felix Gers (porting 2D to CAM-8), the author's Japanese colleague Hemmi and his programmer assistant Yoshikawa (translating CA rules to Boolean expressions), and two M. Sc. students from Nara Institute of Science and Technology (NAIST). At NTT, there were 3-4 people from the Content Addressable Memory machine group who were finding ways to apply their machine to CAM-Brain. So, things are certainly hotting up.

(Note added, May 1996) - Fig. 15 shows about 800 artificial neurons with their axons and dendrites grown using the CAM-8 machine with 128 Mega words of 16 bits. This figure is taken from an 8 square meter poster containing 100,000 neurons. In a year, (using "von Neumannized" one cell wide neural trails) this number will be about *3 million*. Felix Gers thinks he can port the 3D version to the CAM-8. The 3D rules are almost complete and number over 160,000, i.e. nearly 4 million with (24) rotations. Fig. 16 gives a doubly zoomed (by factors of 32 and 50) impression of the scale of the 2D version of CAM-Brain on the CAM-8 machine. Fig. 17 shows CAM-Bryo and CAM-Brain on the CAM-8. (Figs. 15, 16 and 17 were made by my colleague Felix Gers after he ported the 2D growth phase to the CAM-8 machine).

References

[Codd 1968] E.F. Codd, *Cellular Automata,* Academic Press, NY, 1968.

[de Garis 1990] Hugo de Garis, "Genetic Programming: Modular Evolution for Darwin Machines,"*ICNN-90WASH-DC,* (Int. Joint Conf. on Neural Networks), January 1990, Washington DC, USA.

[de Garis 1991] Hugo de Garis, "Genetic Programming", Ch.8 in book *Neural and Intelligent Systems Integration*, ed. Branko Soucek, Wiley, NY, 1991.

[de Garis 1992] Hugo de Garis, "Artificial Embryology : The Genetic Programming of an Artificial Embryo", Ch.14 in book *Dynamic, Genetic, and Chaotic Programming*, ed. Branko Soucek and the IRIS Group, Wiley, NY, 1992.

[de Garis 1993] Hugo de Garis, "Evolvable Hardware : Genetic Programming of a Darwin Machine", in *Artificial Neural Nets and Genetic Algorithms,* R.F. Albrecht, C.R. Reeves, N.C. Steele (eds.), Springer Verlag, NY, 1993.

[de Garis 1994] Hugo de Garis, "Genetic Programming : Evolutionary Approaches to Multistrategy Learning", Ch.21 in book "Machine Learning : A Multistrategy Approach, Vol.4", R.S. Michalski & G. Tecuci (eds), Morgan Kauffman, 1994.

[de Garis 1995] Hugo de Garis, "Cosmism : Nano Electronics and 21st Century Global Ideological Warfare", (to appear in a future nanotech book).

[Drexler 1992] K.E. Drexler, *Nanosystems : Molecular Machinery, Manufacturing and Computation,* Wiley, NY, 1992

[Goldberg 1989] D.E. Goldberg, *Genetic Algorithms in Search, Optimization, and Machine Learning,* Addison-Wesley, Reading, MA, 1989.

[Toffoli & Margolus 1987, 1990] T. Toffoli & N. Margolus, *Cellular Automata Machines*, MIT Press, Cambridge, MA, 1987; and *Cellular Automata Machines*, in Lattice Gas Methods for Partial Differential Equations, SFISISOC, eds. Doolen et al, Addison-Wesley, 1990.

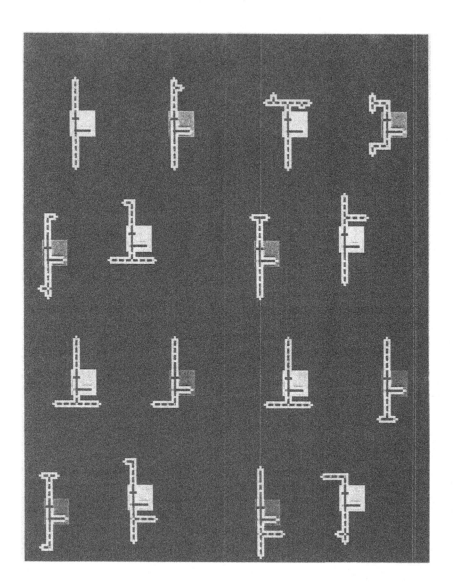

Fig. 9 2D CAM-Brain Early Growth

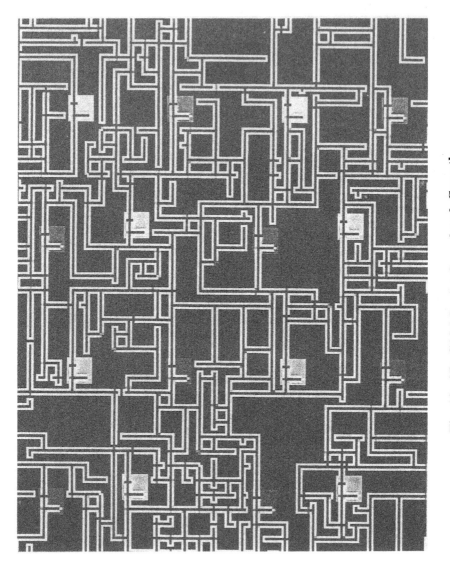

Fig. 10 2D CAM-Brain Completed Growth

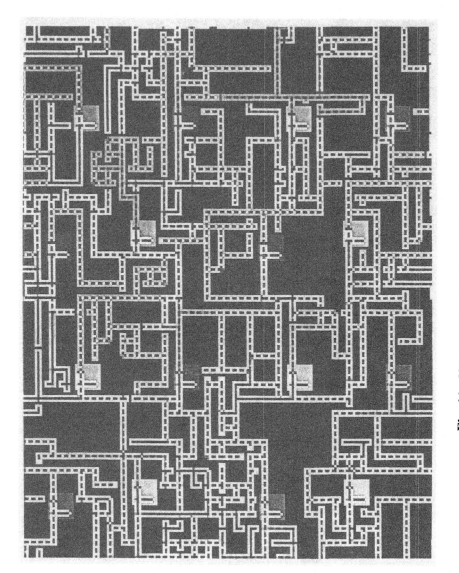

Fig. 11 2D CAM-Brain Neural Signaling

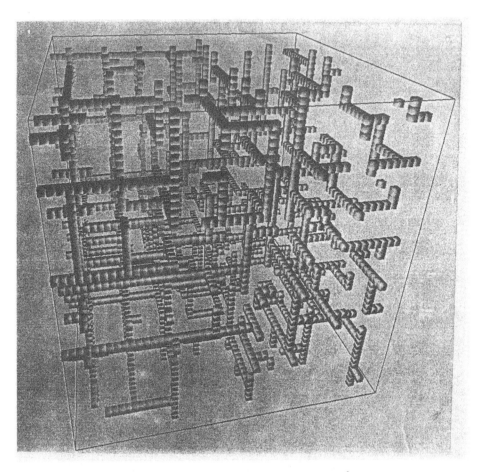

Fig. 12 3D CAM-Brain Non-Synaptic Growth

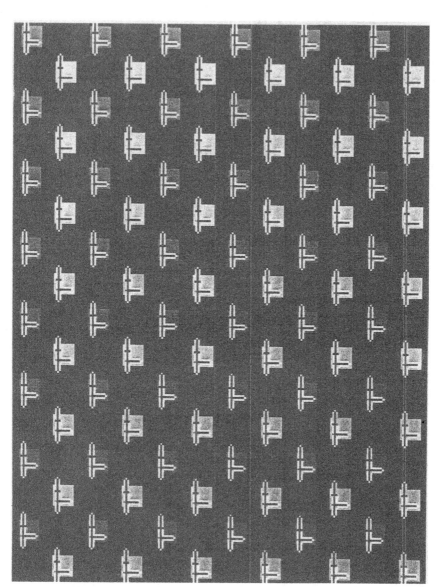

Fig. 13   2D CAM-Bryo

240

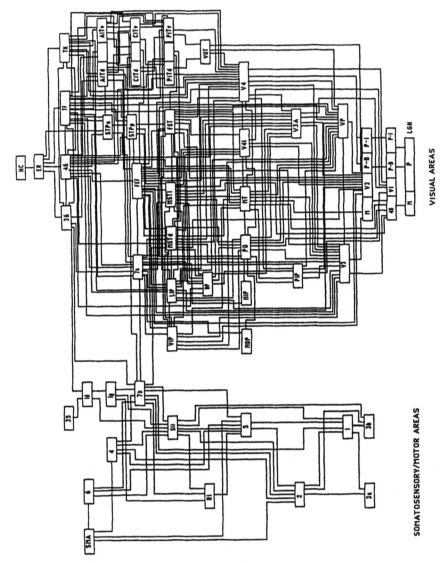

VISUAL AREAS

SOMATOSENSORY/MOTOR AREAS

Fig. 14 van Essen's Monkey Brain Architecture

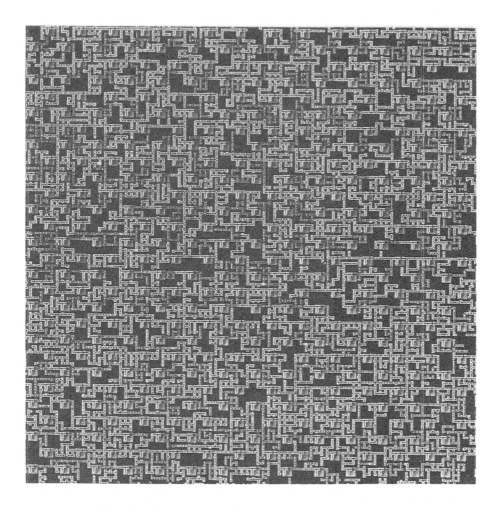

Fig. 15 2D CAM-Brain on MIT's "CAM-8"

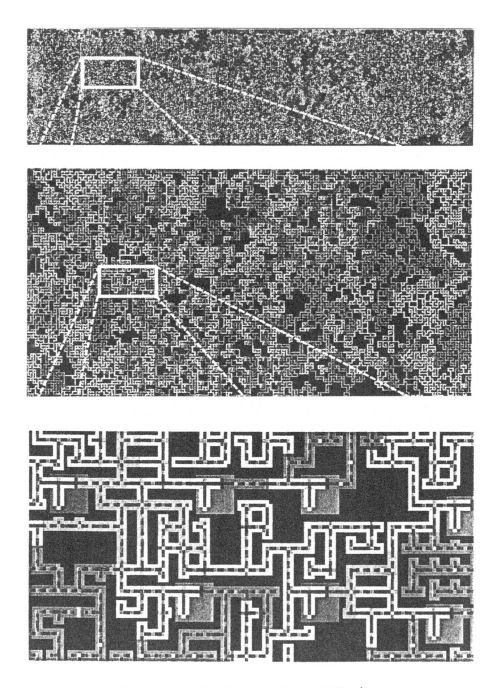

Fig. 16 Doubly Zoomed 2D CAM-Brain

243

Initial Pattern

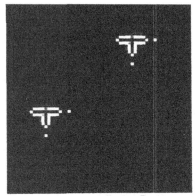

Growth of Neuron Bodies

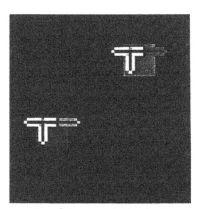

Excitatory and Inhibitory Neurons

Chromosomes

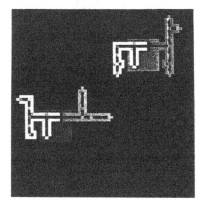

Growth Phase of Axons and Dendrites

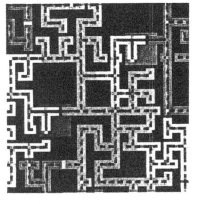

Grown Neural Network

Fig. 17 CAM-Bryo and CAM-Brain

Springer
and the
environment

At Springer we firmly believe that an international science publisher has a special obligation to the environment, and our corporate policies consistently reflect this conviction.

We also expect our business partners – paper mills, printers, packaging manufacturers, etc. – to commit themselves to using materials and production processes that do not harm the environment. The paper in this book is made from low- or no-chlorine pulp and is acid free, in conformance with international standards for paper permanency.

 Springer

Lecture Notes in Artificial Intelligence (LNAI)

Lecture Notes in Computer Science